Unitivity Theory

A Theory of Everything

by

Leroy R. Amunrud

Bloomington, IN

Milton Keynes, UK

AuthorHouse™
1663 Liberty Drive, Suite 200
Bloomington, IN 47403
www.authorhouse.com
Phone: 1-800-839-8640

AuthorHouse™ UK Ltd.
500 Avebury Boulevard
Central Milton Keynes, MK9 2BE
www.authorhouse.co.uk
Phone: 08001974150

© 2007 Leroy R. Amunrud
No part of this book may be reproduced, stored in a retrieval system, or transmitted by any means without the written permission of the author.

First published by AuthorHouse 07/07/07

ISBN: 978-1-4259-9295-8

Printed in the United States of America
Bloomington, Indiana
Library of Congress Control Number: 2007904854

This book is printed on acid-free paper.

This book is dedicated to the memory of
Sir Isaac Newton
who contributed greatly to the
knowledge of light,
gravity and the calculus.

Preface (the recommended place to begin reading.)

This book possesses two approaches to understanding our universe. One approach is to look at the universe from the perspective of theoretical physics. The second approach is to look at the universe from the perspective of axiomatic mathematics. The end product in each case is unitivity theory. Unitivity theory possesses phenomena consistent with the phenomena of our universe, and further, possesses not only a unified field theory, but also, a "unified theory of everything." This "unified theory of everything" covers gravity, magnetism, electricity, light, force-carrying particles, quarks, charges, etc., and shows that all these have a common, unifying base. To the knowledge of the author this is the first book to handle the whole universe in a unified way and to disclose that the universe is as simple as one, two, four. The universe of unitivity theory possesses one activity, two parallel subspaces that constitute its entire space, and four basic types of particles that constitute its entire building blocks. Collectively these three entities constitute the unifying base for everything in the universe. It is because of this unified treatment of the whole universe that the name "unitivity" has been chosen.

This book can be read at two different levels. A person having general knowledge of an automobile can drive it, but if one wants to understand it better, one should be a mechanic. Similarly, any inquisitive person can gain general knowledge of gravity, light, magnetism, charge, and other phenomena by reading and perusing selected parts of this book, but to understand unitivity theory at a deeper level, one should be a physicist or a mathematician and read this book on unitivity theory as an introductory textbook.

When this book is read from front to back, one is not reading the chapters in the order that they were written. Chapter 2 was written in 2000; Chapter 4 was written in 2003; Chapter 10 was completed in 1996; the remaining chapters were written in 2004 and 2005. The main chapters are 2, 4 and 10. Chapter 2 develops the universe as a real model that meets the axiomatic conditions stated in Chapter 10. Chapter 4 develops unitivity using a theoretical physics approach. Chapter 10 is where unitivity theory really had its beginning. This chapter presents a mathematical, formal-axiomatic development of unitivity theory including many theorems with their proofs. Some scientific readers may prefer to start their reading by going to Chapter 10. Others may prefer to start their reading by going to Chapter 2, or Chapter 4. A fan of a sport that uses a bouncing or hit ball should read Chapter 1 and then jump to Chapter 12.

Unitivity theory is a theory that in general is not well known at the time of the writing of this book. Because it contains so many new concepts concerning the universe, these concepts are introduced first in one setting using one set of words, and then reintroduced a number of times in other settings using slightly different words. In this way the reader is exposed to, and becomes familiar with, a more complete picture of the universe. To illustrate this point the following story is related.
Only one part of a given elephant was allowed to be examined by each of five blind persons. They were then instructed to use their findings to describe the whole elephant. The one that felt the tail said, "The elephant is like a short rope." The one that felt the side of the elephant said, "The elephant is like a huge boulder." Another who felt an ear said, "The elephant is like a small, thick rug." The one that felt a leg said, "The elephant is like the trunk of a mighty oak tree." The fifth blind person felt the elephant's trunk and said, "The elephant is like a long tube."
In order to become more familiar with an elephant and have a more complete picture of an elephant, each of these persons needs to be exposed to an elephant from a number of different perspectives. This is the approach taken in this book. Many concepts and properties are presented over and over again, each time from a slightly different setting. If the reader does not clearly understand a property the first time it is presented, they are encouraged to read on. Being exposed to a given property a number of times, and under different surroundings, tends to increase one's understanding of this property.

The author of this book was educated in country schools and a small town school. He did not understand gravity, but was very sure that at the university level gravity was completely understood. Upon graduating from high school, he would have preferred to stay on the ranch, but the ranch was too small for this to be an option. The next-best thing seemed to be to go to college and satisfy a perplexing curiosity he had which was to understand what gravity really is. In college, while studying the findings of Sir Isaac Newton, he was exposed to equations that describe the force of gravity. These findings were very impressive. But, when it came to understanding what gravity really is, the answer given was, "Gravity is force at a distance and warped space 'somehow' produces this force." This was not satisfying. This book is the result of constantly wondering about gravity, magnetism, electricity, and light, and repeatedly employing mathematics, and physics, to reveal the hidden mystery of these and other phenomena of the universe.

Some of the important results obtained in unitivity agree with relativity, but many facets do not. The main reason this is true is that, in every aspect of unitivity, energy and momentum are conserved, while in relativity this is not true. The structure for the universe of unitivity contains matter-energy, dark-matter-energy, and dark-energy in quantities that agree with recently measured percentages. All of these energies are housed in the space of unitivity, and they all have specific roles to play. The space of unitivity is not empty and it is not totally fixed, but rather it is a dynamic space which expands and contracts and carries along each and every mass object contained in it. The reason that mass is carried along is because all mass in its mass state is completely encased in the ether space of unitivity. Unitivity space is a space that has ether present everywhere within it. With the aid of the creative-destructive activity, ether density (warping) produces the magnitude and direction for each and every gravitational force vector, magnetic force vector, and electric force vector. Ether tensile strength provides the traction needed for accelerations. A non-empty, ether space yields a unique velocity for each and every mass object's discrete wave-movement through it. The possession of a creative-destructive activity and an ether space are necessary in order for the universe to conserve energy and momentum.

Surprisingly, when viewed from a housing Euclidean space, the movement of large objects is very much like the relative motion of relativity. Two large objects can be "very nearly" at rest in ether space and yet, because of the dynamic ether space, be moving apart at a "large" velocity relative to each other; again this is as viewed from a housing Euclidean space.

The gravity of unitivity imparts acceleration to the nucleus of an atom similarly to the way acceleration is imparted to an automobile at rest on a flat surface. First, there is a source of existing energy. For the automobile it is gasoline in the tank, and for gravity it is kinetic energy stored inside each and every quark in the form of two circuits of oppositely rotating anti-mass. Next, there is a means of converting the given source of energy into mass kinetic energy. For the automobile this is the operating of a motor with a transmission, and for gravity this is the operating of the creative-destructive activity (or equivalently, the folding-unfolding activity). This activity operates on an electron-positron pair which is in a form called a gyro1. Gyro1s are the building blocks for quarks. Each gyro1 is repeatedly folded and then unfolded. Last, there is traction. For the automobile this means the tires grip the road, and for the quarks in the nucleus of an atom this means the vibrating gyro1s open and grip the ether space. (An automobile is formed from atoms too, and if one goes down to the level of the automobile's gyro1s, unitivity reveals that the acceleration of the automobile requires vibrating gyro1s to open and grip the ether space. Why is this true? The tires gripping the road push the mass in the car's atoms ahead into ether space, but the ether space will not let any mass in a mass state pass through it. The force on the ether space produces a local, density-gradient field in the ether space pointing in the same direction as that of the applied force. Now this new density-gradient field acts similarly to a gravitational, density-gradient field. This newly formed density-gradient field causes the ever vibrating gyro1s to open in the direction of the density gradient. This opening of the gyro1s makes them capable of gripping the ether space and producing acceleration. So, at the gyro1 level, the two cases are identical. This illustrates the close relationship that exists between a field produced by a pushing force and a gravitational field. These properties are repeatedly addressed in this book.)

Similarly, every light wave moves by repeatedly using the creative-destructive activity to operate on pairs of matched electrons and positrons. Each paired electron and positron simultaneously unfold and then simultaneously fold back in a form called, a gyro2. A gyro2 grips the ether space in a similar way to that of a gyro1, but its discrete-type movement is at an average speed that is equal to the speed of light. By contrast, in order for a gyro1 to move at this speed it would have to be completely opened which happens to be impossible. Each of the two types of gyros moves mass discretely in a wave form. Gravity affects a gyro2, but not in the same way or as much as it affects a gyro1. All of these facts are developed in detail at various places throughout this book.

In order to illustrate the way the conservation of energy and the conservation of momentum will be found to permeate all of unitivity theory, the following three examples are presented to the scientifically oriented reader in a very brief outline form.

1. De Broglie waves and light waves are quite similar, yet quite different. For both of these waves, there is a force-carrying particle that is a shaped, quantized, empty hole or flaw in ether space that is produced by the unfolding activity. Each force-carrying particle in connection with the folding-unfolding activity produces a discrete movement of mass through ether space. In de Broglie waves and in light waves the folding and

unfolding of the nuggetrons' energy, simultaneously with a paired amount of pocketons' anti-energy, results in these energies canceling each other. This insures that energy is conserved. Within the creative-destructive activity the associated pull forces and push forces on nuggetrons are offset by equal, but opposite forces on the ether's pocketons. In this way momentum is conserved. This book contains a detailed derivation of these facts.

2. The conservation of energy can be employed to show that, if m represents the amount of mass of the nucleus of an atom at rest, then the associated paired anti-mass moving discretely in its interior circuits is represented by m, also. This follows from the fact that at the gyro2 level, a half reflection of a light ray produces no change in energy, but can convert a gyro2 into a gyro1. This change causes a gyro2's mass, M, moving at the speed 2c for half of time to be converted to a paired amount of anti-mass effectively moving at the speed 2c for half of time. The conservation of energy by the creative-destructive activity requires that M represents the paired amount of anti-mass also, because then the kinetic energy of the mass, namely, M c-squared, is completely converted over to the kinetic energy in the anti-mass, which is now M c-squared, also. The atom's nucleus is formed using gyro1s, thus by heredity it follows that the nucleus of an atom with rest mass, m, contains anti-mass with kinetic energy equal to m c-squared. This illustrates that Einstein's equation E = m c-squared is correct. This book on unitivity theory explains in detail why this equation is true.

3. In de Broglie waves, the original rest mass and the original paired anti-mass contained in the nucleus of each atom are moved in such a way that the sum of their kinetic energies is invariant when measured by a rod and a clock at rest in zero gravity. This means that in any acceleration caused by gravity, energy is conserved in a very interesting way. In order to conserve energy, gravity converts existing anti-mass kinetic energy within a given nucleus into an equivalent amount of mass kinetic energy, and simultaneously, it increasing the unit of time for this falling nucleus. This makes gravity a perpetual motion activity. If this were not the case, the universe's "battery" would have gone dead a long time ago. For mass moved in the wave form of unitivity, the ether space is left in tact, and there is no friction. A "fast" moving atom with its "slow" moving anti-mass in the circuits inside its quarks possesses a larger unit of time than an atom at rest. This book explains all this and shows that an atom's changes in time are consistent with the Lorentz transformations.

This is getting too far ahead of the story. The reader is encouraged to study this book by starting at the beginning and learning unitivity one concept at a time. Unitivity theory establishes that all of the phenomena of the universe have a unifying base that is simple as one, two, four. The establishing of this base and then using this base to produce the phenomena of the universe is what this book is all about.

People who gave assistance and encouragement, asked questions, sent clippings, arranged talks, produced visual aids, did proof readings and experiments in connection with this book include: Dr. Loren Acton, Donald M. Scott, Dr. Chris Steel, Gerald D. Nordley, Arden Amunrud, Arla Amunrud Pribnow, Scott Graber, Mark Wolfe, George Dalthorp, Lois Dalthorp, my wonderful wife Doris and our children Mark, Marcia, Erica, Lisa, and Brent. Thank you to all these people and to all of the instructors who so patiently taught me. I extend an extra special "thank you" to Mark for his help on moving gravitational fields and bouncing steel balls, to Mark Wolfe and Brent for producing the cover on this book and to Erica for her many long hours of hard work producing the manuscript.

CONTENTS

Part One: UNITIVITY FROM A THEORETICAL PHYSICS PERSPECTIVE

Part Two: UNITIVITY FROM A MATHEMATICAL PERSPECTIVE

1 Introduction

On the nature of unitivity

Unitivity is a theory that is based on material axioms that have been tested over and over in the laboratory and have always been found to be true. (Also, in Chapter 10 unitivity is established using formal axioms where there are no imposed conditions on them other than that they must be consistent.) Unitivity is a unique theory in that it starts with empty space and then applies five material axioms to produce a universe that satisfies these axioms from the very beginning. At the time this book is being written, there is no other known theory that applies basic laws during the creation of the universe. The standard practice is to start with a universe that is formed, and then to begin applying the basic laws that govern it.

Unitivity theory reveals when these laws are applied in the very beginning, a more complete structure of the universe is found, and this structure then can be used to replace the partial, incomplete structures that have been used for many centuries. This more complete structure necessitates the introduction of a number of new terms in order to describe it. Even people well versed in physics will have to take time to add some new words to their vocabulary. This is unfortunate, but obviously must take place. Note that the basic laws used to develop unitivity theory are universal and always check out to be true. This implies that any theory that violates these basic laws in its development is, at least in part, a false theory.

What is the structure of the universe? Unitivity reveals the universe is simple as 1, 2, 4. It possesses one activity, two parallel subspaces that constitute its entire space, and four basic types of particles. This structure for the universe is a far cry from the structure arrived at by the Greeks who said the universe consisted of the four basic elements fire, water, earth and air. Nevertheless, the ancient Greek's approach to the universe was on the right track. Unitivity's approach to understanding the phenomena of the universe is basically the same as Euclid's approach to understanding the geometry of the plane.

Observations are important especially for rather "large" objects, but very "small" objects defy being seen by employing objects which are much larger than they are. Also, the axioms for unitivity reveal that there are types of energy that are dark energy and not visible. The material axiomatic system and the formal axiomatic system of unitivity each act like a mirror enabling the universe to look at itself and behold its elegant grandeur including even its dark energies.

Before going into the derivation of unitivity theory, let us jump right into two important revelations of this theory. The chosen revelations are the inner workings of gravity and light. These revelations concerning gravity and light will be explained in some detail, and most likely, in more detail than the reader at this point in their reading can fully understand or appreciate. Actually, it will be impossible to grasp everything presented in this chapter because the new concepts introduced at this point are introduced only in a very brief, outline form with their proofs left out. Also, many words are used that have not been defined. These new concepts are thrown on the table mainly to get the reader started thinking in a new direction. After reading the definitions, explanations and proofs that are presented in other chapters within this book, the reader may find it enlightening to occasionally come back and reread this chapter. In fact, except for the bouncing ball mystery part, it might be more appropriate to place this chapter at the end of the book as a review chapter rather than placing it here as an introductory chapter. But one has to begin introducing the new ideas of unitivity theory in some setting, and the selected one is the following brief overview of unitivity theory's gravity and light.

Creative-destructive activity

Unitivity theory assumes that the universe came from empty space into existence. This demands that a creative activity exists. In order for mass to have a wave form it is obviously necessary that a destructive activity exists. The following axiom is taken to be true in the material axioms for our universe:

There presently exists a creative-destructive activity that repeatedly forces basic mass particles into existence in quantized, circular loops out of empty space, and alternately operates in reverse and forces each and every one of the created basic mass particles back out of existence in such a way that, if at any time all the energy and anti-energy currently in existence are combined, they completely cancel one another.

As previously stated the full development of the structure of the universe will not be done at this point; rather just a few things that are pertinent to light and gravity will be presented in a brief overview form.

During the creative phase the creative-destructive activity pushes on nuggetrons, the building blocks for mass using an offsetting force on pocketons, the building blocks of ether space, and during the destructive phase it pulls on nuggetrons using an offsetting force on pocketons. The magnitude of the pushing and pulling forces are each proportional to the local density of the pocketons that fill ether space, and each of the forces on the nuggetrons tends to be in the direction of the pocketons' density gradient vector in the local, involved ether space.

The creative-destructive activity repeatedly acts on all mass causing it to alternate between a folding or mass phase and an unfolding or wave phase, with an equal amount of time being spent in each of these phases. There are two, basic, mass forms that are building blocks and are involved in this vibrating process. One of these mass forms is the gyro1 which is composed of an electron and a paired positron that are side by side "four dimensional doughnuts" in 4-space, but coexist as one "three dimensional doughnut" in 3-space. The other is a gyro2 that also is composed of an electron and a paired positron, but its form in 3-space is a pair of tangential "doughnuts" with their maximum, exterior circles in a common plane. The gyro1s are "small" magnets and are the building blocks for quarks. The gyro2s in their maximum mass state are visible light, and in their completely destroyed, wave state they are photons. In an electromagnetic wave, a gyro2 in its maximum mass state is always at rest and encased in ether space, and it is moved at twice the speed of light during the time its associated photon is being formed.

Gravitational field

A gravitational field will be defined when a unified field theory is developed. For now let a gravitational field be described as existing in an ether space that is composed of pocketons of two types where the pocketon's density gradient vectors in each of two parallel, 4-dimensional subparts of ether space are non-zero and identical at each 3-dimensional point within the gravitational field.

Gravity revealed

Gravity is the creative-destructive activity in operation as it produces forces by repeatedly unfolding and folding every bit of mass that exists within the given gravitational field.

Gravity produces a force on a mass object by producing a force on each and every nuggetron in the object's make-up. This force is produced during both the unfolding phase and the folding phase, and this force tends to be in the direction of the common density gradients within the gravitational field's two parallel subspaces. At all times half of the nuggetrons that make up a single, at-rest quark are in the mass or folded state and half are in the wave or unfolded state, and this ratio is continually maintained. This literally means that each quark is really a pair of half-quarks. At the precise time all the involved gyro1s are switching from a folding phase to an unfolding phase or vice versa, there is a half-quark that exists in a mass state, and there is a paired half-quark, which contains an equal number of nuggetrons, and it is in a wave state. (It is established later why at the time of switching half of the gyro1s making up a quark are in a mass state and half are in a wave state.) In quarks, as each nuggetron in the mass state is unfolded it simultaneously is replaced with a folded nuggetron that has just returned from its wave state. Thus, the number of nuggetrons in the mass state for a given quark at rest in ether space is invariant. All nuggetrons associated with a quark are treated equally, and each one is pulled out of existence and each one is pushed back into existence. If a quark is composed of n nuggetrons that exist in the mass state, then 2n nuggetrons are involved in the weight of this quark during each and every complete cycle of unfolding and folding. This illustrates that for the nucleus of an at-rest atom, there is a relation between its mass and its total weight, but the number of nuggetrons involved in the total weight is twice the number of nuggetrons that makeup the nucleus's mass state.

The movement of a gyro1 through ether space takes place in its wave phase. Two out-of-sync sets of gyro1s walk a quark through ether space as the two sets alternate between their mass state and their wave state. Under accelerations, adjustments are made in the amount of mass and the time required for each vibration. This is discussed in Chapter 10 when the Lorentz transformations are discussed. This description of the quarks that make-up any associated nucleus is so very important that is repeated below using slightly different words.

The nuclei of atoms

All nuggetrons, at all times, are blinking in and out of existence at a very large number of frames per second. This is the way gravity continually determines the precise amount of mass that exist in each and every mass object, and further this is the way mass objects are moved. Mass in a folding phase is completely at rest in the local ether space, and in an unfolding phase it is moved a "small" discrete step as a de Broglie wave at a velocity that is equal to twice its observed velocity.

For the nucleus of any atom at rest in ether space, half of its mass exists in a folded form and half of its mass in an unfolded, wave form. In addition, nuggetrons and associated pocketons are being folded and added to the nucleus at a fixed rate, and at the same time, nuggetrons and associated pocketons are being unfolded and removed from the nucleus at this same rate. Each and every nuggetron in the nucleus is equally involved. This means that the total number of nuggetrons, that are in the mass state and make-up the nucleus of any at-rest atom, is invariant.

The nucleus has a weight that is produced when each and everyone of its nuggetrons is folded and when each and everyone of its nuggetrons is unfolded. For a nucleus at a given fixed point in a given fixed gravitational field, the total number of nuggetrons being folded or unfolded per unit of time is invariant. Thus, it follows that at this same point the total weight of this at-rest nucleus is constant.

When there is acceleration, the number of nuggetrons and pocketons being unfolded during each vibration must increase in order for the atom to move through ether space. Mass in its mass state is completely encased and is at rest in ether space, and further it is carried with any movement of the ether space that surrounds it. An increase in mass is produced in any gyro1's movement through ether space, and in order to conserve energy, there also has to be an increase in the time used to complete each vibration of the involved gyro1. Exactly what happens under accelerations requires mathematical calculations and these calculations are given in Chapter 10, theorem 16 when the Lorentz transformations are produced.

Discrete movement of mass

The frequency for blinking mass is a very "large" number and the discrete movements are normally very "short" distances. This makes the movement of any atom approximately continuous, at least as long as its velocity is "small" compared to the velocity of light. The number of frames per second is so large, that at "low" velocities, all mass appears to move continuously. This, along with the added smoothing that is produced by the walking type of movement of the gyro1s in quarks, makes it permissible to use the calculus when studying physics.

At a very large velocity, the nucleus of an atom becomes unstable due to the slowing down of the anti-mass moving in the two circuits that hold quarks together, and due to the increase in the amount of anti-mass that is getting mixed in with these circuits due to the movement. This mixing causes the time for each vibration to increase and the length of each discrete step to increase. This is true because in each of the gyro1s that make up the nucleus of an atom, there is only one folding-unfolding chamber for each of its two circuits of anti-mass. Each of these chambers must continually unfold a fixed amount of anti-mass per vibration that is contained in one of the atom's circuits, and simultaneously for movement, unfold additional anti-mass of this same type in front of the atom. Next this chamber must refold the fixed amount of ant-mass back into this same circuit and also, refold additional anti-mass behind the atom. The amount of anti-mass folded and unfolded outside the internal circuits is increased when the atom is accelerated, and also, under acceleration the discrete step length for each gyro1 is increased. This puts extra strain on the anti-mass circuits. For this reason atoms become unstable when their velocity is "greatly" increased.

Light enlightened

A gyro2 in its maximum-mass state is light.

Light for any single light ray is repeatedly at rest in ether space and has the form of a quantized, rounded figure 8 when viewed from the side, and when viewed from the front, the form of a line segment (I) having a quantized length, a quantized width and a unique angle of rotation. This is the light one sees.

Illuminated light enlightened

Illuminated light is the creative-destructive activity repeatedly moving a gyro2 through ether space as a photon which is a shaped, quantized empty hole in ether space that is produced when the given gyro2 is destroyed, and which, in turn, is filled back when the given gyro2 is reformed at its leading end.

This means that illuminated light exists, alternately, in two different quantized states. It exists as an at-rest gyro2, which is mass in the form of a quantized rounded figure 8, and then it exists in its wave state as a photon which is a side-by-side, matched pair of identical, quantized, right-circular-cylinder holes in ether space having a common length that is equal to the given ray's half-wave length.

Note that illuminated light is a moving gyro2 that is alternating between its rounded figure 8, mass form and its "double barreled shotgun" laying down wave form.

The following is the trail of a light ray moving to the right. The pictured wave part or photon part is in reality two laying down side-by-side right circular cylinders that go from one rounded figure 8 to the next rounded figure 8 and are crosswise perpendicular to each of the figure 8's.

S N
\+ pole - pole +

Illuminated Light: 8∿∿8∿∿8∿

\- N + S -
pole pole

In a light ray the mass, while in the folding phase, is completely at rest in the local ether space and, in the unfolding phase, is moved at a linear velocity equal to twice the average speed of light. (This is as measured by a local rod and clock at rest in zero gravity.) The unfolding produces a photon which is a matched pair of quantized, right-circular-cylinder holes (or flaws) in ether space. For convenience purposes this matched pair of side-by-side holes in three-space will usually be referred to as a single, right-circular-cylinder hole having two subparts. The length of a photon is half the wave length of the associated light ray. This means that the discrete movement of light through the ether space always has the average speed, c. This is one form of the constancy of the speed of light in unitivity theory. Note that in a light ray the electric and magnetic fields are alternating and are rotated at a right angle with respect to one another.

Color is detectable in a light ray's mass state. In this maximum, mass state, the paired electron and positron are momentarily completely folded, and are completely at rest in the local ether space. This means the color of light is determinable by detecting the mark that is left by the light ray's maximum mass size. The "doughnut" electron and "doughnut" positron in their maximum folded state for red light are thinner and have a longer circumference than they have in blue light. This seems backwards, but in Chapter10 it is established that this fact produces the equation: "the quantized energy of a given light ray is equal to its frequency times Planck's constant."

Perpetual motion activity

A startling revelation of unitivity is that because, the creative-destructive activity conserves energy, it is a perpetual motion activity. This makes gravity and light perpetual motion activities. However, if any one of the activity's folding-unfolding chambers is destroyed, then it follows that this chamber ceases to be a perpetual motion machine, and the associated force-carrying particle is now a stranded, quantized, shaped, empty hole in ether space, which is simply heat. This means that entropy is increased whenever a folding-unfolding chamber is destroyed.

Creative-destructive activity exists, but what is it?

Most readers at this point have a logical question in their mind. Gravity, atoms and light all employ the creative-destructive activity. The logical question is, "What is the creative-destructive activity?" The mathematical answer is that it is known to exist and many of its properties are known. However, its inner make-up is left open. Its existence can be stated in an axiom, because the universe is presumed to have come from empty space and it is now here. Thus, the creative activity must exist. Some of the main properties of the creative-destructive activity are the following: It is known that this activity uses a counter force on a folding pocketon to produce a push on an associated folding nuggetron of the same type, and uses a counter force on an unfolding pocketon to produce a pull on an associated nuggetron of the same type. Its associated linear unfolding rate is 2c for half of the time. The shaped, quantized holes it forms in the sea of ether pocketons are the force-carrying particles. Exactly what the inner make-up of this creative-destructive activity happens to be is left open in unitivity theory.

To a mathematician, the important thing is to know that the creative-destructive activity exists and to know its usable properties; it would be nice, but it is not imperative, to know its inner make-up. The number one in the Peano axioms is undefined and therefore open, but its properties are given and these properties are usable. People have done arithmetic successfully for centuries not knowing exactly what the number one is.

A physicist can study the properties of such things as light, atoms and gravity "without knowing exactly what they are", but by the same token a physicist can study and understand gravity, atoms and light much better "by knowing very nearly what they are", even if this knowledge is based on an open term. Properties of building blocks influence the phenomena of the universe as much, or possibly more, than knowing exactly how the building blocks are made. The determining of exactly what the creative-destructive activity happens to be is a topic that more properly comes under the heading of theology or philosophy.

Press on

Having been exposed to part of unitivity theory in a very brief overview the reader is encouraged to read on, with increased determination, to learn about other phenomena of the universe and to fill in the details of the structure of the universe according to unitivity. To eventually learn such things as how velocity and momentum of an atom are related to the velocity and momentum of a light ray by examining the way the two use force-carrying particles; to learn how light is reflected, polarized, colored, etc.; to learn the nature of charges, molecules, electricity, magnetism and related items. Unitivity is a new approach to the whole universe. Thus, the reader must exercise patience in learning this theory.

Bouncing and hit-ball mystery introduced

Some books contain a mystery that is not solved until the reader reaches the very end of the book. The mystery for this book is the questions, "How do balls bounce?" or "Why is it easier to hit a home run when hitting a fast pitched ball than when hitting a slow pitched ball?" To illustrate this mystery, let a steel ball be suspended on a string and let another steel ball having about four times the weight of the first steel ball be suspended on a string from approximately the same point. Pull back the larger steel ball a circular (or better an inverted cycloid) distance x and then let it go, allowing it to strike the first steel ball which is at rest. Next pull both of the balls a circular distance x in opposite directions away from their at rest position, and simultaneously let them go. When they hit squarely the small steel ball bounces back about twice as far in the second case as it did in the first case. In the second case one would think the larger steel ball would be slowed down as it brings the incoming small steel ball to rest. This would leave it with less driving power to bounce the small steel ball away than it had in the first case. How do you explain that just the opposite happens? The small steel ball moves away from the larger steel ball much faster in the second case than it does in the first case. If you think the expansion of a denting-in can explain this, note that in the case of steel balls, if there is much of a denting-in, it will remain in the balls, it will be visible, and it will not produce much of a bounce. Let this unsolved mystery be called, "the bouncing and hit-ball mystery."

For discussion:

1. Is there any problem with the equation E = m c-squared and the conservation of energy in the "big bang"?
2. How does gravity conserve momentum?
3. Explain the difference between mass and weight. Is the exact number of nuggetrons that make-up a given object equal to the exact number of nuggetrons associated with its weight? Hint: In one sense they are equal in another sense the number of nuggetrons that make-up the given object is half the number of nuggetrons that are involved in the object's weight.

2 A Real Model for the Formal Axiomatic System

In this chapter real laws possessed by our universe are used to establish that the universe we live in is a real model for the formal axiomatic system as given in Chapter 10. To do this one must have a procedure for establishing the existence and structure of real objects in our universe which in turn can be used to define the undefined terms in the formal axiomatic system. This means that each object that is shown to exist, also, must be shown to possess the same properties as those possessed by the undefined term it replaces. Lastly, it will be noted that when a given material axiomatic system truly and completely defines our universe, and when a given derivation using these axioms contains no errors, then the derived structure is the true structure of our universe.

Axiomatic systems

It is advantageous to start with a brief introduction to axiomatic systems and how they are used. There are two basic types of axiomatic systems. They can be described as follows:

1. Material Axiomatic system – Observed laws of the real universe are stated as axioms.

An example is to state the law of action and reaction as an axiom.

2. Formal Axiomatic system – Undefined terms are used. The statements made about these undefined terms are taken to be true within this system of statements. These statements, that are taken to be true, are called axioms. The system of true statements is called a formal axiomatic system.

An example is Peano's axioms for the natural numbers. In this formal axiomatic system, number is undefined. The first axiom states that one is a number that is unique in that it is not the successor of any other number in the family of numbers. In Peano's axiomatic system a person must be sure to distinguish between a symbol for one, i.e. 1 or I, and the "unique number one." A person can write many different 1's or I's, but they all represent the same unique number "one."

The strong point of "material" axioms is that they do have observations to support their validity. One weak point concerning material axioms is that these axioms require observations which sometimes could be impossible to obtain. By contrast a strong point for "formal" axiomatic systems is that they can be used to study undetectable, abstract things by hypothesizing their existence and then stating their properties.

Axiomatic example

To illustrate how we are going to use formal axiomatic systems and material axiomatic systems let us consider the three stages an inventor goes through in order to come up with a new invention. To begin with, the inventor has an idea in the back of his mind. This idea envisions a unique object that never has existed, but possibly could be produced and used to perform some special task. He then decides what properties this object is to possess. The next step is the designing stage. In this stage he considers the properties of available materials and pertinent physical laws. He then draws blueprint diagrams stating how this object is to be made in order for it to have the desired properties. The final stage is that of taking the blueprint diagrams and building the sought-after object.

The beginning stage is equivalent to writing a formal axiomatic system. One considers an undefined object that never has been seen. One then writes statements giving the components and true properties that this object is to possess. The second stage is equivalent to writing a material axiomatic system. In this stage one considers available materials and pertinent laws, and lists the roles they are to play in order to produce the sought-after object. In the final stage the desired object is made and checked to determine whether or not it has the right properties to make it a real model for the material axiomatic system and a true model for the formal axiomatic system.

Example of a formal axiomatic system

These points are illustrated in the following example. Henry Ford had a dream. He wanted to build a good horseless-carriage. In the back of his mind he envisioned a model T Ford that he could drive. Let us put this dream into a formal axiomatic system.

> Formal Axiomatic System for a Horseless-Carriage:
> There exists a horseless-carriage (undefined) which has the following properties:
>
> axiom 1. It has a front part and a following part.
>
> axiom 2. It has a transmission (undefined) that can be used in conjunction with a motor (undefined) to make a horseless-carriage self-propel itself ahead over a given surface.

axiom 3. It has a steering wheel (undefined) that enables one to turn a moving horseless-carriage.

axiom 4. It has a brake (undefined) that can be used to halt a moving horseless-carriage.

Any object, that satisfies all of these axioms, is called a model for this formal axiomatic system. I will not attempt to go into the material axiomatic system that Henry Ford used to design his horseless-carriage, but among other things he would have used the law of action and reaction, a detailed description of the motor, transmission, steering wheel, brakes, etc. Note that a model T Ford is a real model for both the given formal axiomatic system and the referred to material axiomatic system.

Related story containing an important point

An interesting story is told about Henry Ford's first horseless-carriage. When he had finished his first prototype, he took it out for a test drive. It was working really better than he had expected, so he went a little further than he had planned. He turned down a lane to showoff his horseless-carriage to a farmer using his team of horses in a nearby field. But, before he had gone very far, the lane dead-ended. He tried to prove that being capable of stopping and turning while going ahead, was equivalent to the property of being capable of going in a reverse direction. Try as he might, he could not get out of the narrow lane. He finally concluded that being capable of backing-up is an independent property that can't be obtained using the properties that his present model T ford possessed. He reluctantly asked the farmer with the team to pull him backwards out of the lane. You can guess the rest of the story. He went home and in some form or other added axiom 5 to the axiomatic system. The added axiom 5 is:

axiom 5. It has a transmission (undefined) that can be used in conjunction with a motor (undefined) to make a horseless-carriage propel itself backwards.

As a system, the first four axioms can all be true and axiom 5 can be either true or false. This proves that axiom 5 is an independent axiom in this system. It should be noted that adding axiom 5 makes axiom 4 a dependent axiom. A property of the real numbers states you cannot go from going ahead (at a positive velocity, v) to going in reverse (at a negative velocity, $-v$) without having a velocity zero at some point in time.

The main thing to learn from this example is that an independent axiom (or property) cannot be obtained from the other axioms (or properties) in the system. Henry Ford could not produce a reverse movement using just the first four axioms. It is impossible to accomplish this task, because having a reverse is an independent property. Automobiles and standard motorcycles both satisfy the first four axioms. By contrast, automobiles satisfy axiom 5, but standard motorcycles do not satisfy axiom 5. This proves that axiom 5 is an independent axiom.

If one is trying to find a unified field theory or any other fact concerning the universe, and one relevant, independent axiom (or property) that defines our universe is missing then one is attempting the impossible. One will never obtain the desired result. This is so very important that to emphasize this point, the following story is related.

A missing independent axiom leads to attempting an impossible task

A fellow was walking down a very dark street. He took out his light-up-face watch to see what time it was. The watch just flipped out of his hands. He then proceeded down the block to a street light on the corner. Getting down on his knees, he started looking for his watch. A second person came along and asked him if he had lost something. He replied that he had lost his watch. They agreed it was big enough that they should be able to find it. After looking for some time without finding the watch, the second gentleman asked the first gentleman if he had any idea where he was standing when he lost it. He replied, "Back in the middle of this block." He then added, "It was too dark back there to ever find it. So, I came to this street light so that I could see enough to find it."

The moral of the story is, "If the watch is not there, then you are not going to find it there." By the same token, one cannot obtain a unified field theory, if a relevant, independent axiom that is required in the definition of our universe is missing. The reason that this is true is that a unified field theory does not even exist in an axiomatic system when a relevant independent axiom that it is based upon is missing. If it is not there, then you are not going to find it there. This point has been stated twice to emphasize its importance. Attempting a task where a required independent axiom is missing, is just like trying to put a jig-saw puzzle together when some pieces are missing. One is never going to be able to finish this impossible task.

Formal axiomatic systems

Formal axiomatic systems are good for looking in the dark because you do not need to know what an object looks like. One is mainly interested in the properties possessed by undefined objects. This is very advantageous when working in our universe with particles that are of the type associated with the Heisenberg's uncertainty principle or are of the type that puts them in the dark-energy category.

The formal axiomatic system given in Chapter 10 produces events that occur in a manner consistent with our universe. However, there are a number of undefined terms that need to be defined in order to obtain a real model for this axiomatic system. This is accomplished by using a material axiomatic system to define our universe, and then employing this material axiomatic system to derive the basic structure of our universe, and finally to establish that the obtained universe is truly a real model for the formal axiomatic system. The material axiomatic system that is used for defining our universe is the following:

Material axiomatic system that defines our universe

axiom 1. For every action there is, simultaneously, an equal reaction or anti-action in the sense that if the two are combined, there is no action.

axiom 2. For every change in energy there is, simultaneously, an equivalent change in anti-energy in the sense that if the two are combined, there is no change in total energy.

axiom 3. Two different particles or energies cannot coexist, but they may appear to coexist. (Specifically what is true in unitivity theory is that two different, 4-dimensional, particles or energies cannot coexist in 4-dimensional space, but when they are of different types, they may appear to coexist in a 3-dimensional subspace of 4-space.)

axiom 4. There presently exists a creative-destructive activity that repeatedly forces basic mass particles into existence in quantized, circular loops out of empty space, and alternately operates in reverse and forces each and every one of these just formed basic mass particles back out of existence in such a way that if at any time all the energy and anti-energy currently in existence are combined, they always cancel one another.

axiom 5. Our universe is made as elementary as possible.

Remarks on these axioms

Let us discuss the very important axiom 4. One is confronted with the fact that gravity continues to produce force over extremely long intervals of time without running out of energy or even weakening. What is the nature of a process that can go on indefinitely without consuming any total energy? What makes gravity a perpetual-motion machine?

Similarly, what propels an electromagnetic wave for long extended journeys through ether space? It carries a very small amount of energy and yet uses this energy very efficiently.

These observations indicate that in our universe there is a process by which energy can be harnessed without changing the total energy of the complete system. One very simple way, and possibly the only way, to do this is to have a process that goes on repeatedly, and makes either equal increases in both energy and anti-energy where, anti-energy is a canceling type energy, or makes equal decreases in both energy and anti-energy. It is obvious that in both of these cases, the sum, total energy does not change, and yet there is a change in the amount of energy of each type.

It turns out that this same process is needed in order to make it possible for mass to take on a wave form that in turn gives it the capability of moving through ether space without creating a flow of ether around itself. One very simple and again possibly the only way to accomplish this feat for mass and ether space whenever they cannot coexist, is to have mass moved in a destroyed form (wave form) through a destroyed part of ether space and then be restored at a new forward position in the restored ether space. The quantized circular loops mentioned here are called spin-orbits in the formal axiomatic system for our universe where they are shown to be a workable configuration for producing this type of movement. (The symmetry in action and reaction is observed to be of the highest order when, in 3-space, a wheel of coexisting positive and negative nuggetrons are rotating oppositely in each of the gyro1s that makeup a given quark, and a wheel of nearly-coexisting positive and negative circuit pocketons are rotating oppositely through the axis of each of the involved gyro1s. This configuration is a remarkable "wheel in the middle of a wheel.") One may ask, "Are there other workable configurations?" The answer to this question is left open in unitivity theory. The above observations concerning the creative-destructive activity are used to establish the right to insert axiom 4 into the material axiomatic system and apply it to our universe.

This axiom really helps in the development of unitivity theory. Unitivity theory establishes that the forces associated with this activity's directional pushing of basic mass particles into existence is the source of all momentum including that of light, and the canceling of this directional pushing requires a crossover-in-direction pulling which in turn is the source of all movement of mass as waves.

This very unique activity is the "motor" that gives gravity the capability of accelerating mass objects. The equation, $F = ma$, is universally accepted. It is well known that gravity accelerates mass objects, and further, gravity obviously is part of our universe. Thus, gravity must possess a "motor" that produces the force F. The activity introduced here is the essential "motor" for the great mystery called gravity. Within any given quark, this unique activity produces potential kinetic energy in the form of circuits of rotating anti-mass that gravity can draw upon to impart kinetic energy to the given quark, and by heredity to all mass objects made from quarks. Furthermore, it can do this without producing a change in the total energy of the universe which always has the value zero.

For axiom 3, the reason a 4-dimensional space is required is made clear when density gradient fields in the two parallel and independent subparts of space are discussed. Squashing is needed in order to produce an increase in density in each of these two, parallel, independent subspaces. One should note that in "relativity's" approach to the universe, time is referred to as the fourth dimension, but for "unitivity" a fourth dimensional space variable is required.

Why is axiom 5 needed? It is needed in order to obtain a unique structure for our universe. If one can accomplish a task, one can always find a different, more complicated way to accomplish this same task. If one workable structure is found for our universe, and then if more things are added to this structure in such a way that a workable structure is maintained, how can one determine which of these two structures is the true structure of our universe? According to axiom 5 the more elementary one is the true one.

Axioms 1 and 2 are universally accepted and require no further comments.

Our universe as a real model

To obtain our universe as a real model, we start with empty space and with material axiom 4. That is, we start with zero energy and the activity given in material axiom 4. Our universe is known to possess mass-energy particles. What do the given material axioms disclose about our universe's most basic particles? When one uses axiom 4 to create just one most-basic, discrete, mass-energy particle, one ends in failure because axiom 1 and axiom 2 are both violated. When one tries to create or form just two most-basic, discrete, mass-energy particles, axiom 1 demands that the second particle has to be made exactly backwards, in every sense, from the first particle. This second particle has to be an anti-particle of the first particle. Now axiom 1 is satisfied, but axiom 2 is not satisfied because the energies of these two mass-energy particles are formed alike except for direction and therefore, their energies are additive. Note that kinetic energy does not depend on the direction of the force that produced it. In order to comply with both axiom 1 and axiom 2 it is necessary that two additional basic anti-mass-energy particles be created exactly backwards from each other and be formed simultaneously with the first two basic mass-energy particles. This establishes that four types of discrete basic particles or discrete basic energies are required, and they must come in foursomes.

Consider the four basic types of particles as displayed in figure 1.

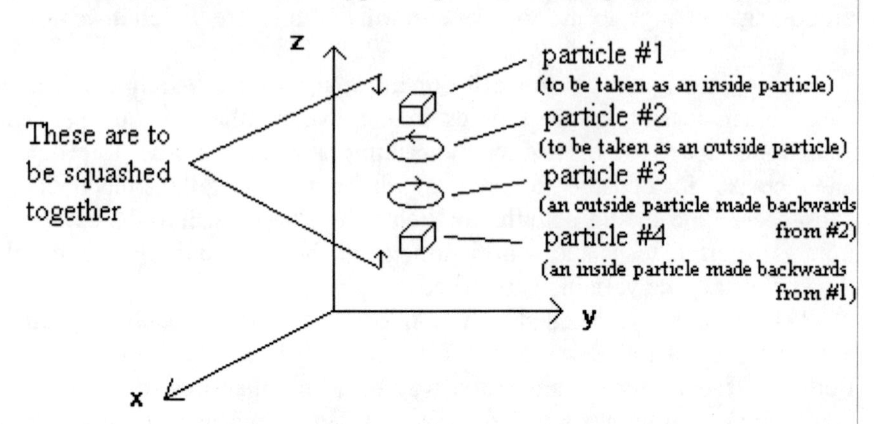

Figure 1. Four particles produced preserving energy and satisfying the law of action and reaction.

Each of the four types of most basic particles will now be described. The descriptions verify that these four types of basic particles satisfy axioms one and two. Later on these four types of basic particles are given names. Also, mass-energy is often referred to as just mass or just energy, and anti-mass-energy is often referred to as just anti-energy or just anti-mass.

Particle #1 – is an anti-energy particle, a positive space particle, an anti-mass particle for mass particle #3, and the action particle for reaction particle #3 (and #4).

Particle #2 – is an energy particle, a negative mass particle, an anti-particle for particle #3, and the action particle for reaction particle #4 (and #3).

Particle #3 – is an energy particle, a positive mass particle, an anti-particle for particle #2, and the reaction particle for action particle #1 (and #2).

Particle #4 – is an anti-energy particle, a negative space particle, an anti-mass particle for mass particle #2, and the reaction particle for action particle #2 (and #1).

Note that there are two types of energy that makeup total energy, namely energy and anti-energy. They are created in equivalent amounts and have the capability of canceling one another. Also, there are two types of basic mass (or energy) particles with each type possessing an associated basic anti-mass (or anti-energy) particle. To obtain the quantized energy for a light ray in our universe, (See axiom 5 in the formal axiomatic system of Chapter 10.), the magnitude of the energy in each of these discrete, most-basic mass particles must be equivalent to the energy 3.31×10^{-27} gm cm^2/sec^2. This number is half of Planck's constant.

Four basic types of real particles

By axiom 5, the number of most-basic particle types used to form our real universe should not exceed four. When four will do, to use more would make things more complicated than is necessary. It is assumed that our universe is defined by the given material axioms. Thus, it must be as elementary as possible, and therefore, our universe must be formed from four different types of most-basic particles. The existence of our universe insures the existence of these four types of most-basic particles. These four types of real particles can be used to define the undefined terms in formal axioms 2 and 3 as given in Chapter 10. This is done as follows.

Let

particle #1 be the pocket filled with positive space, and let it be called a positive pocketon;
particle #2 be the nugget of negative mass, and let it be called a negative nuggetron;
particle #3 be the nugget of positive mass and let it be called a positive nuggetron;
particle #4 be the pocket filled with negative space, and let it be called a negative pocketon.

The required property that pocketons pull together, and the required property that pocketons fill all of our universe's ether space in two separate, but adjacent, parallel subparts, will be discussed next. The reason two separate subparts are necessary is presented first and the necessary pulling together aspect is presented last.

The necessary two separate subparts for space

How many separate parts must space have in order to contain the four types of real particles? One cannot put particle #1 with particle #3. Particle #3 is a positive mass particle and #1 is its anti-mass particle. To put them in the same part of space would be like putting the gingham dog in with the calico cat. According to the Chinese plate, they would eat each other up. This same event is literally true for mass placed with its anti-mass. They combine and both go back to empty space. Similarly, particle #2 cannot be put together with particle #4. Thus, one needs at least two separate subparts for the space of our universe.

It will be proved that the ether space of our universe only needs to have two subparts. Just place particles #1 and #2 together in one subpart of our space and place particles #3 and #4 in the adjacent, but separate, subpart of our space. Now axiom 5 implies that two subparts is the unique number of subparts for the ether space of our universe because two subparts is as elementary a structure as is possible. Two subparts for the ether space of our universe are necessary and two subparts are sufficient.

It is encouraging that formal axiom 3 in the formal axiomatic system of Chapter 10 requires two subparts in ether space, and four different types of particles. This makes it easy to define these undefined terms in the formal axiomatic system. But the big question is, "How are the real pocketons to be placed in real space so that they meet the "separate" or "independent" condition that is required for the two subparts of space as stated in formal axiom 3 in Chapter 10?"

The formal axiom 3 states that space is composed of two intermingled, but distinct, finite sets of pockets. One of these sets has all of its pockets filled with negative-space (These are the negative pocketons.) and the other has all of its pockets filled with positive space (These are the positive pocketons.). Parts 7(a) and 7(b) state that the two types of pockets must be placed in such a way that they can carry two separate density gradient fields. Again, one logically asks, "How is this possible?"

A two dimensional example

To make this easier to explain, we first will solve the problem for an apparent two dimensional, universe. In this case it is easy to draw the diagrams needed to illustrate the way this is accomplished.

Consider the following subspace of Euclidean three-space. Let x and y be non-negative numbers where $x \le a$, $y \le b$ and magnitude of $z \le \varepsilon(d)$ with $a > 0$, $b > 0$, $\varepsilon(d) > 0$ and $\varepsilon(d)$ is the length of one side of a cube that constitutes the region of influence of a given pocketon having density, d. Note that ε is a function of the pocketon's density, d. See figure 2.

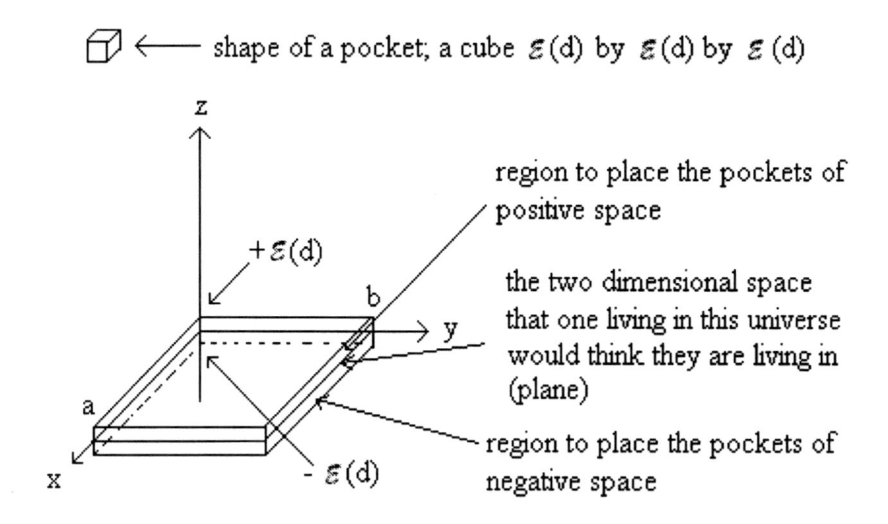

Figure 2. The required two parts of space for axiom 3 for an apparent two dimensional universe.

Note that the above two separate, independent spaces are adjacent and parallel and can be filled with cubes. In some cases these cubes may become distorted, but locally they always extend approximately the same distance in the *x*, *y*, and *z* directions. We have satisfied axiom 3 for this special case, when we place the negative pocketons filled with negative space into the region between the planes given by $z = -\varepsilon(d)$ and $z=0$ with $0 \leq x \leq a$, $0 \leq y \leq b$, and the positive pocketons composed of positive space between the plane given by $z=0$ and $z=\varepsilon(d)$ with $0 \leq x \leq a$, $0 \leq y \leq b$. This is exactly what one needs to carry two separate, independent density gradient fields and formal axiom 3 is satisfied in this special case.

For this apparent two dimensional situation, a person living in this universe would think they were living in a plane, when in reality, they are living in a three dimensional subspace of Euclidean three-space. Because $\varepsilon(d)$ tends to be a very small positive number, they are "almost" living in a plane. It should be noted that trying to tell an intelligent person living in this plane that there exists a third dimension, would leave the person bewildered. He would understand only two orthogonal directions in 2-space, and could not envision or find a direction for a possible third orthogonal direction.

A three dimensional universe possessing string points

To satisfy formal axiom 3 in an apparent three dimensional space, one starts with a four dimensional space. See figure 3.

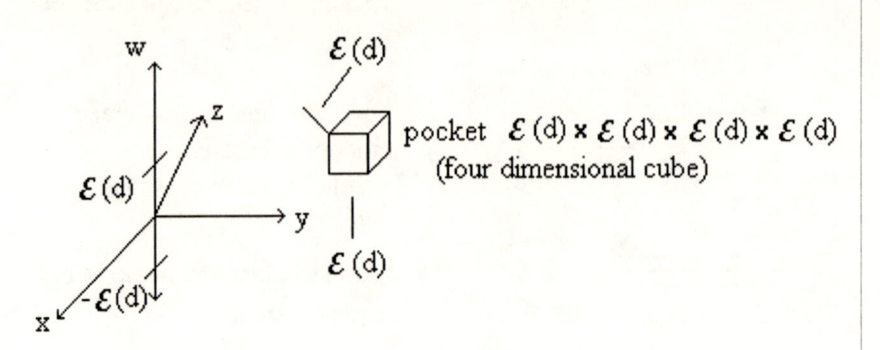

Figure 3. Pockets in four dimensional space and pocket's shape.

In unitivity theory the four base vectors for four-space are always assumed to be selected so that they are orthogonal in the sense that every possible pair of vectors in the basis for four-space have an inner product that is equal to zero. One now uses *w* just like *z* was used above. Place the four dimensional cubic pocketons filled with negative space between $w = -\varepsilon(d)$ and $w=0$. Place the pocketons filled with positive space between $w=0$ and $w=\varepsilon(d)$. In each case, the placing of the pocketons leaves no space between the cubic pocketon's regions of influence. Adjoining any point $(x, y, z, 0)$ of this particular three dimensional subspace of Euclidean four-space, there is both a pocketon filled with negative space and a pocketon filled with positive space. To show this, go to any point $(x, y, z, 0)$ in the described ether space that is filled with pocketons and move toward the point $(x, y, z, -\varepsilon(d))$. One is moving in or on the surface of a pocketon filled with negative space. Similarly, go to this same point $(x, y, z, 0)$ where the ether space is filled with pocketons and move toward the point $(x, y, z, \varepsilon(d))$. One is then moving in or on the surface of a pocketon filled with positive space.

In terms of physics, the points that are observed to be three dimensional in our universe are actually, one dimensional, short, line segments (open strings) in four-space. Note that we live and make observations "between" the two subparts of space, and the unseen fourth dimensional variable is normally extremely "small" in absolute value. This is to say that we only see the common starting point of the two independent "short" line segments that makeup a string point. Thus, practically without exception, one only needs to use the first three-space variables when computing distances. Formal axiom 3 of the formal axiomatic system now is satisfied by our real universe, that appears to be three dimensional, but in reality it is a four dimensional universe possessing an ether space which has two parallel, basically-independent subspaces as described above. These two subspaces are basically independent, however they must adjust and keep working toward having an ether space where the positive and negative pocketons are in a 1-1 arrangement as described in the formal axiomatic system in Chapter 10. This is where magnetism gets involved.

Properties of pocketons

Let us now discuss some features of the pocketons. These are anti-mass particles. Thus, one should expect them to act backwards from mass particles. When a pocketon is compressed, if there is to be an indication that this has occurred, it must respond by either pushing or pulling harder on the surface of all adjoining pocketons of its type. At this point one must employ observed properties of gravity to determine which of these properties pocketons possess.

If a compressed pocketon pushes out like compressed mass pushes out, then the compression is weakened as the adjacent pocketon being pushed on moves away. This is true even when this movement is restricted to be along a single straight line. When the pocketons respond to a compression with a pulling force on its adjacent pocketons, then as adjacent pocketons move in, staying in rank, the compression is supported. In order to obtain a stable density pattern, a signal must be sent out from any given compressed pocketon delineating how much it has been compressed. To accomplish this one must impose that the greater the difference of the density of two adjacent pocketons, the proportionately greater is the pull per square unit between them. This means that a newly compressed pocketon causes adjoining pocketons, of the same type, to move toward this newly compressed pocketon and in turn causes these adjacent pocketon to have an increase in their density. In this way it is possible for our universe to expand and contract and have its space warped.

The way a neutral-mass sun informs a neutral-mass earth that it is out there, and vice versa, is for each to compress its neighboring pocketons of each type equally which in turn instigates, respectively, a pull on the two types of pocketons. The pull instigated by the neutral sun produces two identical density gradient fields of pocketons in the two subparts of space with each pointing toward the sun. The pull on the pocketons instigated by the neutral earth produces a second pair of identical density gradient fields in the two subparts of space each pointing toward the earth. The resulting combined density gradient field at each point in each of the two subparts of space is the vector sum of two density gradient vectors at that point. (Later it is established that this gradient density field is a gravitational field with the magnitude of the creative-destructive activity's pushes and pulls per unit of mass being determined by the density of the pocketons at a given point, and the direction of these pushes and pulls being determined by the density gradient vectors in the pocketons at this same point. Our goal is to eventually establish that this is the way a gravitational field is produced.)

Pocketon-density waves

A change in density of pocketons is an event of just expanding or just contracting. It is shown later that light requires an unfolding followed by an at-rest folding in order to move. This indicates that pocketon-density waves (gravitational waves) are faster than light waves. This in turn means that the pocketon-density waves are "extremely" fast and therefore rather hard to detect per se. The changes in pocketon-density, and the adjustments to the changes in density, may be close to being instantaneous, but because the pocketons are pliable, these adjustments cannot be instantaneous. When one pulls or pushes on a rod with no give (an inelastic rod), the pull or push is transmitted instantaneously, but when there is give (a pliable rod), then there is a wave and a delay.

Density gradient fields

Given that the additional pull per square unit is directly proportional to the additional difference in density between adjacent pocketons, what is the nature of the density gradient field that is formed when a point source of neutral mass is introduced into the ether pocketons? The following calculations are made in a housing Euclidean space.

Let the more dense pocketons be basically a source-point density with the pulling influence moving out in all directions equally. The volume of influence, v, in a three dimensional setting is given by

$$v = \frac{4}{3}\pi r^3 .$$

The change in volume as a function of r, is given by

$$\frac{dv}{dr} = 4\pi r^2 .$$

With increasing r, this surface area increases like r^2. The law of action and reaction implies that the pull per square unit between adjacent pocketons is dropping off like $1/r^2$. This in turn implies that the density of pocketons is also dropping off like $1/r^2$ because the change in pulling force per square unit between adjacent pocketons is directly proportional to their change in density.

From the formal axiomatic system for an ether type universe (axiom 5, part 6 in Chapter 10), the folding and unfolding process tends to push or to pull the associated mass in the same direction as the direction of increasing density with a force per unit of mass that is proportional to the density of the pockets in that small neighborhood. This now means that for the formal axiomatic system, the force of gravity per unit of mass drops off like it is supposed to drop off in that it drops off like $1/r^2$ whenever the density of the pocketons drops off like $1/r^2$. (This point is discussed in Chapter 4, also.)

Notice that the pocketons when squashed locally near one point, upon reaching stability, form a density gradient field that drops-off like one over r-squared. Assuming a spherically-shaped, ether-space universe, this explains why the universe is expanding whenever the density of pocketons on the outer edge of the universe is dropping-off at a rate greater than one over r-squared. When the dropping-off rate becomes a one over r squared rate, stability is attained. If the dropping-off rate is less than a one over r-squared rate, the universe will be contracting or collapsing until a one over r-squared rate is attained.

The activity given in material axiom 4 forces the four real particle types into existence. This activity in reverse then logically forces or pulls them out of existence. It was noted above that one must require the force produced between any real pocketon and real nuggetron involved with the creative-destructive activity to be directly proportional to the density of the local pocketons and nuggetrons. This yields a real process involving real pocketons and real nuggetrons that is in complete agreement with formal axiom 5, part 6. When coupled with the way mass moves as a de Broglie wave, this is basically the reason the formula for the force of gravity in our real universe has a distance squared in its denominator. It follows from the fact that for a given source-point of neutral mass that is locally compressing pocketons, the density of the pocketons drops-off like $1/r^2$.

Gravitational formula for a two dimensional universe

It is interesting to determine the formula for the force of gravity for an apparent two dimensional universe. On a plane, the area of influence, when starting at a source-point, is a circle. For a circle with a radius r, the area is given by

$$A = \pi r^2.$$

The rate of change of area with respect to a change in r is given by

$$\frac{dA}{dr} = 2\pi r.$$

Going through the arguments as above, the density is dropping off like $1/r$. In this case the formula for gravity would have an r to the first power in its denominator.

Gravitational formula for a one dimensional universe

It is even more interesting to determine the formula for the force of gravity for an apparent one dimensional universe. Length ℓ on a line replaces volume or area as given in the above calculations; i.e. one has

$$\ell = r$$

$$\frac{d\ell}{dr} = 1.$$

This implies there would be a one in the denominator of the formula for the force of gravity in this case. This means the force of gravity is constant in an apparent one dimensional universe.

Unified field theory

The structure for our universe is now complete enough to carry a unified field theory. By placing negative nuggetrons in the subspace of positive pocketons at one 3-dimensional point (region) of our visible universe and by placing a number of positive nuggetrons in the subspace of negative pocketons at a different 3-dimensional point (region), two density gradient fields are formed. The reason density gradient fields are formed is due to axiom 3. Neither nuggetrons and/or pocketons can coexist in 4-space, thus squashing must take place when they are all forced into the same box. This squashing of the pocketons causes them to pull together to form a density gradient field in each of the two adjacent, but separate subparts of space. The combined, overall field formed here is an electric field. It is explained later how positive and negative particles move correctly in this field.

When an equivalent number of positive and negative nuggetrons are placed at the same 3-dimensional point (region), the density gradients in the two subspaces at any given point are identical, and the resulting combined field is a gravitational field. Note that this is equivalent to placing a point source of neutral mass at a 3-dimensinal point (small region).

When extra positive pocketons are added at a 3-dimensional point (region) in the subspace composed of positive pocketons and the same number of extra negative pocketons are added at another 3-dimensional point (region) in the subspace composed of negative pocketons, the field formed is a magnetic field. This field is a density gradient field similar to an electric field, but it dissipates with increasing time because the pocketons nudge themselves back into their 1-1 equilibrium state. This nudging phenomenon is explained below.

Note that this is a true unified field theory because there is only one field, which is composed of two independent density gradient subfields in the two seas of pocketons (the two parallel subparts of ether space). The formulae for gravity, electricity and magnetism all depend on the manner in which the ether pocketon's density drops off in this one field. This insures that all of these formulae for force in a source-point situation will have an *r*-squared in their denominators. The field is an electric, magnetic, or gravitational field, depending on what particles are used to produce the field and also, on how these particles are located with respect to one another.

Nudging process

The nudging phenomenon for the pocketons is described in a special setting which makes it easier to illustrate. The phenomenon is explained in an apparent two-dimensional universe where the pocketons are three dimensional cubes. In our apparent three-dimensional universe the nudging takes place similarly.

Given that each, of a group of pocketons of either type, is matched face-to-face and 1-1 with the pocketons adjoining it, there is no nudging and further there is no need for nudging. The pocketons are capable of reaching a stable state by merely expanding and/or contracting in a maintained, fixed arrangement.

Consider the case where two pocketons of one type are placed into a slot that is supposed to be occupied by only one pocketon of this type. Assume here that all of the pocketons are of the same type with the pocketons to the left being less dense than those to the right. See figure 4.

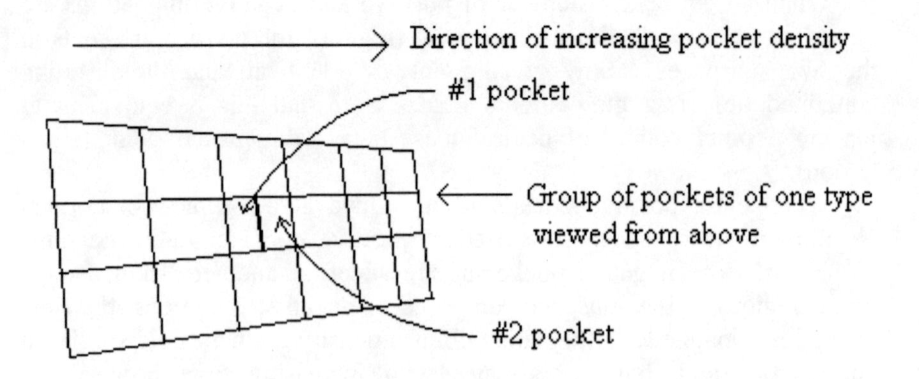

Figure 4. Pockets viewed from above with one overloaded slot.

Label the two pocketons in the overloaded slot, #1, on the left and, #2, on the right. There is a greater difference in density between #1 and its neighbor to the left than the difference between #2 and its neighbor to the right. Recall, the change in the pull per square unit between pocketons is directly proportional to the difference in their densities. Thus, pocketon #1 will move in the direction of decreasing density causing a chain reaction type movement (a movement like a train moves) with the next pocketon moving to the left, etc., until such time that some pocketon finds an open slot. This real nudging process can now be used to define the undefined process in formal axiom 3, part 8 of the formal axiomatic system.

Density gradient field and nudging

It is important to point out that when the #2 pocketon above is replaced with a properly charged nuggetron, then there is no nudging, rather this results in a squashed pocketon which in turn pulls adjoining pocketons in toward itself. This is the way a density gradient field is produced. This is the main difference between an electric field and a magnetic field. An electric field is produced using nuggetrons which in turn is stable. A magnetic field is produced using pocketons which in turn means that there is nudging and the field must constantly be supplied with additional pocketons in order to be maintained. When iron filings are placed over a magnet one can see the lines of nudging and they are of two types. Positive pocketons are being nudged away from a south pole and eventually, toward a north pole and negative pocketons are being nudged away from a north pole and eventually, toward a south pole. The places where iron filings line up are along the lines where the nudging is taking place. This is where the density gradient vectors have the greatest magnitude.

Spin-orbits defined to be quantized loops

The properties of negative mass particles and positive mass particles will be discussed next. In our universe electrons and positrons and other charged particles are known to exist. It has been established that the only mass building blocks that exist in our universe are the positive and negative nuggetrons. Thus to keep things as elementary as possible, it must be that electrons (or negative quantized loops) are composed of negative nuggetrons and positrons (or positive quantized loops) are composed of positive nuggetrons. How is this accomplished?

Figure 1 is consistent with formal axiom 5 in the formal axiomatic system of Chapter 10, provided the production of foursome groups of basic real particles is repeated many times over with the produced group of negative-mass particles being sewed into a real negative, quantized loop (to be used to define a neg-spin-orbit) and the matched produced group of positive-mass particles being sewed oppositely into a real positive, quantized loop (to be used to define a pos-spin-orbit). These two, oppositely rotating, real quantized loops are used to define the undefined terms in formal axiom 5. As stated above, one example of a real, negative, quantized loop is the electron and one example of a positive, quantized loop is the positron. Note that because positive, quantized loops exist in one subpart of 4-dimensional space and negative, quantized loops exist in the other subpart of 4-dimensional space they can be paired, side-by-side in 4-dimensional space, and then they will appear to coexist in 3-dimensional space. Also, note that in order to have a wave property these charged particles need to continually oscillate between their folded (mass) state and their unfolded (wave) state. Any end nuggetron on a sewed together negative quantized loop is still in contact with anti-mass and unfolding begins instantaneously when the folding ends. (It is a remote possible that this switching is not instantaneous. If it is not, then blue light would be a little slower moving than red light. Reports indicate that this is presently being checked by NASA. Unitivity theory can accept a less than instantaneous switching with really no changes to the theory. But the change from folding to unfolding or vice versa must be extremely quick in order for light to carry colors that are not out of focus.)

Gyro1s and gyro2s (Sometimes they are called gyros1 and gyros2.)

In our universe it is well known that electrons and positrons are anti-particles and that they attract one another. What is formed when they get together? In Chapter 10, Definition 20 two, paired, coexisting in 3-dimensional space, spin-orbits are defined to be an I-gyro-1. When this abstract term is defined using the real quantized loops of the material axiomatic system for our universe, the electron coexisting with the positron becomes a real I-gyro-1 and is called a gyro1. When they are paired tangentially, and possess a common plane, the real I-gyro-2 is called a gyro2. The detailed properties of I-gyros-1 and I-gyros-2 are discussed in Chapter 10. Formal axiom 5, 6 and 7 and some following theorems give the fundamental properties these real gyro1s and real gyro2s must have in order to produce phenomena consistent with our universe. To give them different basic properties than ascribed in these axioms and theorems will in general produce phenomena that are not consistent with our universe.

Swirls and quarks

The only remaining undefined term is found in formal axiom 8, and it is the term, "swirl." The major property of a swirl is that it consists of a number of alternately, completely out of sync gyro1s in close proximity.

A gyro1 consists of a negative, quantized loop (usually an electron) and a paired positive, quantized loop (usually a positron) with the property that during unfolding and folding respectively they pull and push negative pocketons in a direction perpendicular to their common face, and through their coexisting center region, and pull and push positive pocketons in the opposite direction. When these gyro1s are arranged close together, they share pocketons. This means one of the quantized loops of one gyro1 is unfolding pocketons at the same time that an adjacent quantized loop of the same type is folding pocketons and sending them toward the quantized loop that is unfolding. When a group of these out of sync gyro1s is arranged in a distorted circular array, then one has in essence a circular magnet which is a real swirl. Figure 5 shows three joined swirls arranged as a three-leaf rose.

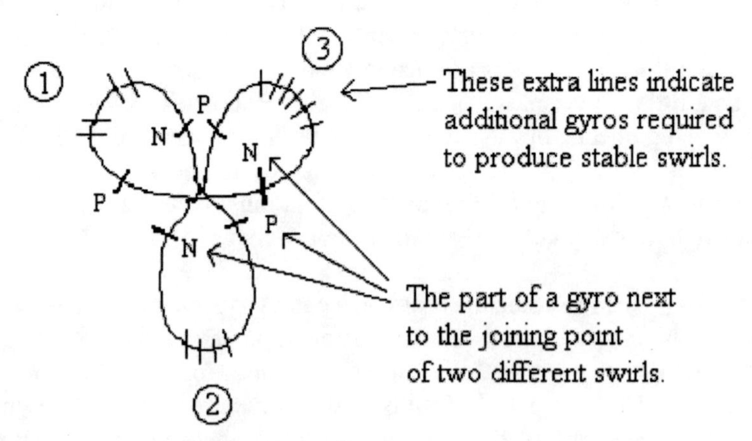

Figure 5. Three joined swirls.

The way a swirl is held together is as follows: The gyro1s must always be circularly arranged in such a way that their positive quantized loops have the same spin orientation when one goes from gyro1 to gyro1 along the circular arrangement. This is necessary in order to keep negative-circuit pocketons rotating in one direction and positive-circuit pocketons rotating oppositely in an angular-momentum-canceling fashion. This shows that a swirl is held together by employing the gluon and two ropes of pocketons which pull on each other with a strong force, or if you prefer, pull on each other with <u>the</u> strong force. The gyro1s are made from real nuggetrons, and the circuit pocketons are real. Thus, these swirls are real entities in our universe and can be used to replace the undefined, abstract swirls in formal axiom 8. The gluon is discussed when force-carrying particles are introduced.

For a three-leaf rose arrangement of three swirls, there are three joining points with a positive, quantized loop on one side and a negative, quantized loop on the other side at each joining point. Note that the gyros1 are composed of two quantized loops, which for the most part are an electron and a positron, but these quantized loops are modified when a swirl is formed. See formal axiom 8, part 3. At each of the three joining points that are in close proximity, either just part of an electron, or just part of a positron, can be shifted to the adjoining swirl, but not a part of both at the same point at the time. In figure 5, if ⅔ of the positron at point P on swirl (2) is slid across to swirl (1) and ⅓ of the electron at N on swirl (2) is slid across to swirl (3) (Note that these are weak force types of movement.), then the charges on the resulting swirls, or quarks are as follows:

		charge	type
swirl (1)	⅔ of a positron picked up	+ ⅔	*A*
swirl (2)	⅔ of a positron lost	- ⅓	*A′*
	⅓ of an electron lost		
swirl (3)	⅓ of an electron picked up	- ⅓	*B′*
A second way is:			
swirl (1)	⅓ of an electron picked up	- ⅓	*B′*
swirl (2)	⅔ of an electron lost	+ ⅔	*B*
swirl (3)	⅔ of an electron picked up	- ⅓	*B′*
	⅓ of an electron lost		
A third way is:			
swirl (1)	⅓ of a positron picked up	+ ⅔	*C*
	⅓ of an electron lost		
swirl (2)	⅓ of an electron picked up	- ⅓	*B′*
swirl (3)	⅓ of a positron lost	- ⅓	*C′*

The above process produced possible arrangements for tying swirls (or quarks) together. It is interesting to note that the ropes (or trains) of the two different types of circuit pocketons, that rotate in opposite directions in the two parts of space, are what hold swirls together. The harder one pulls on a circuit pocketon, of either type, the larger in volume it must become. Consequently, the less dense it becomes. From the property of pocketons given earlier, this means its adjacent circuit pocketons will pull harder per square unit on it trying to get it back to their size. From experimentation, scientists have found that this is precisely the way the strong force works. As stated above, the role of the force-carrying particle called the gluon is discussed when force-carrying particles are introduced.

Our universe as a real model is established

At this point, we have in Chapter 10 a formal axiomatic system where phenomena within this system are consistent with our universe. We have established that our universe is a real model for the given formal axiomatic system. This means that our real universe can be studied by studying the abstract universe as defined by the formal axiomatic system given in Chapter 10. This is an interesting and powerful way to study our universe.

Accept unitivity or reject unitivity?

If a reader does not think unitivity theory's structure for our universe is correct, what must he or she do to justify such a conclusion? There are only a few ways to disprove unitivity theory. The possible ways are outlined here.

Unitivity theory's structure for our universe is based on material axioms. If it can be established that one of these material axioms is not true for our universe, then it is justifiable to disagree with the structure of our universe as given by unitivity theory. The material axioms used to define our universe, with the exception of axiom 4 and 5, are tested billions and billions of times each minute by various activities taking place within our universe, and to the knowledge of the author there is no known case where any one of these laws failed even once. Having said all this, if one counter example can be found proving that one of the first three material axioms is false, one can justifiably reject unitivity theory.

If it can be established that the stated material axiomatic system for our universe is missing a basic independent axiom, then unitivity theory would have to allow the insertion of this axiom, and possibly would have to be modified.

Axiom 5 implies that our universe is as elementary as possible. If someone can produce a workable structure for our universe that is more elementary than the 1,2,4 structure of unitivity then one can justifiably reject unitivity theory. Is this possible? Let the 1 be discussed first. When there is an activity, there is no smaller number of activities than having just one activity. For gravity to maintain a constant force on a given mass object that is located at a fixed point in a gravitational field, this unique activity must exist and stay active. Given a creative activity the most elementary way to keep it active is to run the creative activity backwards whenever mass and anti-mass come in contact. This makes the process spontaneous and as elementary as possible. True this activity does have two sub-activities. One sub-activity involves positive nuggetrons and positive pocketons and the other sub-activity involves negative nuggetrons and negative pocketons. But in order to have charged particles obey the laws of gravity, electricity and magnetism these two sub-activities must exist. This makes it hard to come up with a more elementary activity than that of unitivity theory. With no activity, the way gravity really works can be explained only by using something that resembles black magic.

Now consider the 2 parallel subparts of space. Recall that two subparts to space is required in order to be able to keep mass away from its anti-mass or, if you prefer, energy away from its anti-energy. Thus, to get away from a space having two subparts requires getting away from the four basic types of particles. However, the conservation of energy and action and reaction demand that there be four types of basic particles. Having said all this, if someone can produce a workable structure for our universe that has a one part space and/or uses less than four types of most-basic particles with energy always being conserved and actions always being offset, then one can justifiably disagree with unitivity theory. A one-part space has been studied for millenniums with out producing a unified field theory. The reader who still thinks this is possible should note that unitivity theory definitely implies that this is impossible.

One can refuse to accept axiom 5 and proceed to find a structure for the universe that is more complicated than that given by unitivity theory, and on this basis reject unitivity theory. If this approach is taken, the structure of our universe can never be determined, because there would be thousands of workable solutions with no way of determining which one is the correct one.

Axiom 4 can be questioned, provided one does not accept that there exists a creative-destructive activity that produced our universe out of empty space. The second law of thermodynamics implies that the universe must have a beginning and gravity implies there must be some sort of energy conserving activity. In unitivity theory axiom 4 is an independent axiom that is absolutely necessary in order to derive a workable, elementary structure for our universe. It is highly questionable that a workable structure for our universe can be obtained with out employing this axiom 4. Again, this has been attempted for millenniums without success.

If there is an error in unitivity's derivation of the structure or properties for our universe, then it is justifiable to disagree with unitivity theory. Observed properties of our universe are used to obtain some of the properties of particles. If one of these observations, or one implication of these observations, can be shown to be inconsistent with our universe, then one can justifiably reject unitivity theory. But in this case, one should remove the identified error by merely introducing a modification into unitivity theory.

In conclusion, it should be noted that a workable basic structure for our universe has been discovered. The workable structure is simple as 1-2-4. Given that the defining material axiomatic system for our universe is correct, and given that the succeeding derivation is correct, then one has to conclude that the derived structure for our universe is also correct.

For discussion:

1. How does one define a structure?
2. In unitivity what unifies all fields?
3. In an apparent n-space universe with $n > 0$, what power would r be raised to in the denominator of the formula for the force of gravity?

3 Mathematics: The Magic Mirror for Viewing Our Universe

The definition of mathematics given here not only states what mathematics is, but also, gives information on the ways that mathematics can be used to gain knowledge about abstract and real structures. This definition is the author's own definition of mathematics and is the one used to derive unitivity theory.

Mathematics defined

In order to define mathematics it is necessary to make three relative statements first.

1. Any statement that is taken to be true with no questions asked is called an axiom.
2. A non-empty set of axioms, all of which must be true at the same time, is called an axiomatic system.
3. Each consistent axiomatic system defines a real or an abstract structure.

Now mathematical truth and mathematics are defined as follows:

A. The truth contained in orchestrated (i.e. working together) real and abstract structures plus all extended truth as revealed by the art of reason is called mathematical truth.
B. Mathematics is the art of expanding and organizing mathematical truth into a consistent language with many symbols.

Aspects of mathematics

There are three basic aspects of today's mathematics.

1. One basic aspect of mathematics is to add new structures harmoniously into mathematical truth.
2. A second basic aspect is to discover new, logically-extended, mathematical truth.
3. A third basic aspect is to incorporate these new mathematical truths into a consistent language that agrees with all other accepted mathematical truth.

Fact and theoretical truth

There are two ways to introduce a structure into mathematical truth. First, one can start with an axiomatic system. If this system contains a consistent set of axioms, then it can be used to define and introduce a new structure. Note that the set of axioms must be checked to be sure the axioms do not conflict with one another. They must not imply a fact and at the same time imply the negation of this same fact. Whenever this happens there is no model for this inconsistent set of axioms. However, when the defining set of axioms is consistent, and the axioms it contains do not conflict with known mathematical truth, the associated new structure, and any correctly derived extended truth concerning this structure, constitute new mathematical truth. New mathematical truth in this case is considered to be fact and not fiction.

A second way to introduce a structure into mathematics is to start with a given, real structure, like our universe, and proceed to determine a consistent material axiomatic system that defines it. This is harder. One may think they have found a defining set of axioms for a given real structure whenever a large number of extended truth's phenomena agree with the given real structures phenomena, but in reality this set of axioms still may be "only close" to a definition. As soon as it can be established that there exists one property of the structure that does not agree with the defining set of axioms, the arrived at definition is proved to be false. For this reason any extended truth obtained from this type of defined real structure is referred to as theoretical truth.

This does not mean that theoretical truth that is based on material axioms of definition cannot be trusted. When one is highly confident that the defining axioms are reliable, then properly derived extended truth is equally highly reliable. The material axioms used in Chapter 2 to define our universe have been tested repeatedly and always found to be reliable.

The reliability of unitivity theory is supported also, by the large number of things that come for free in its derivation. Mathematical probability theory indicates that the more things that come for free in the derivation of any theory, the greater is the probability the theory is correct.

Degrees of freedom and things for free

In the derivation of unitivity theory, there are very few degrees of freedom open to be used for finding a workable structure for our universe. For example after obtaining a unified field theory it is imperative that there be a source that produces movement of the pocketons in order to produce a north pole and a south pole. Without such a movement of the two types of pocketons, it would be impossible to form a magnetic field. There exist paired electrons and positrons that are side by side in 4-space. These stable pairs are called gyro1s, and they are defined in such a way that they use the creative-destructive activity to produce the required movements of pocketons, and further do this in an action and reaction conserving manner.

What are some of the things that come free with the defined gyro1s? Some examples are magnets, quarks, Lorentz transformations, gluons, gravitons, bosons, de Broglie waves, etc. Additionally, a gyro1 can be transformed into a stable gyro2 without any change in total energy. Light comes with the gyro2s. For free, with light comes the proper reflection of light, the proper bending of light, the proper color of light, the proper fields for electromagnetic waves, the Doppler effect, the polarization of light, the momentum of light, etc. When so many things come for free, one must conclude that the structure for our universe as given by unitivity is definitely on the right track. One just cannot be so lucky that all these things come out correctly by coincidence only. This type of "lucking out" occurs over and over again in unitivity theory.

One impressive illustration of obtaining something for free is the discovery that gravity produces accelerations without consuming or producing any energy. The way mathematics reveals this property of gravity is discussed below.

Creative-destructive activity

The creative-destructive activity is an activity that operates by producing equal amounts of energy and anti-energy which when combined go back to no energy. This creative-destructive activity takes place in a folding-unfolding chamber. The rate at which energy and anti-energy are folded or unfolded depends on the cross-sectional area of the folding chamber. The associated linear unfolding rate for half of time is twice the speed of light when measured using a local rod and clock that are at rest in zero gravity. For a given folding-unfolding chamber, the number of nuggetrons and the number of paired pocketons that are folded or unfolded per unit of time are equal. (I.e. in a given folding-unfolding chamber the events of folding and unfolding take place at the same rate.) The force on the nuggetrons and the anti-force on the pocketons cancel at all times and the magnitude of these forces is proportional to the density of the local pocketons. When the creative-destructive activity forms quantized loop particles at a rate of n nuggetrons paired with n pocketons per second for a time of t seconds, then the larger the values of n and/or t, the greater is the mass of the formed quantized loop particle, and the greater is the amount of its associated anti-mass. However, there are commensurate conditions that must be met and these conditions put limits on both the size of n and the magnitude of t.

One may ask, "How is the energy associated with the creative-destructive activity quantified?" The apparent kinetic energy associated with the creative-destructive activity in a gyro2 is quantified by noting that in a light ray the mass of a gyro2 is moved at the velocity $2c$ for half of time. If the mass of the gyro2 is M, then its effective kinetic energy is, M times half of ($2c$)-squared for half of time, and the gyro2 is at rest with no kinetic energy for the other half of time. Thus, in a gyro2 with mass M the creative-destructive activity maintains an apparent mass kinetic energy equal to M c-squared. An energy conserving, half reflection, takes a gyro2 and makes it into a gyro1. The amount of anti-mass that is associated with the given mass M is also equal to M, and in the new gyro1 the anti-mass M is moving at the effective velocity $2c$ for half of time, and it is at rest for the other half of time. This shows that the creative-destructive activity in an at-rest gyro1 with mass M produces an apparent, anti-mass, kinetic energy equal to M c-squared which is equal to the apparent, mass kinetic energy of its associated gyro2.

Appearance and reality

Why are the words "apparent" and "effective" sometimes placed in front of kinetic energy? This is done to emphasize that the mass kinetic energy produced by the creative-destructive activity is a discrete type of kinetic energy, and it only appears to be kinetic energy of the ordinary, continuously moving type. All mass is moved discretely as a wave. When mass is folded using the creative-destructive activity it is folded at rest in ether space. Thus, during the folding phase mass has no moving kinetic energy. During the unfolding phase, the unfolding takes place at the linear rate of $2c$, and at the end of this phase the associated mass begins forming at the leading end of the associated force-carrying particle. This indicates the mass has been moved the length of the force-carrying particle as a wave. The force-carrying particle is an at-rest, quantized, shaped, empty hole in space and this hole has no moving kinetic energy either. Thus, the creative-destructive activity moves mass discretely as a de Broglie wave which only has "apparent" or "effective" kinetic energy. But this does not hinder one from quantifying the creative-destructive activity by using ordinary, calculated kinetic energy. (A similar statement holds for effective momentum.)

The step size of the discrete movements is normally very small and the frequency of alternating is so very large that the movement of neutral mass appears to be continuous. Further, the unfolding process is continually moving ahead when it alternates between the two out-of-sync sets of gyro1s in each and every quark. This is like a person walking. The movement of each foot is discrete, but the body moves ahead continuously.

The walking type movement of neutral mass in its congealed form is quite continuous and calculus can be used in its physics. The way things "appear" to be happening and the way they "really" are happening are sometimes quite different. All mass in its gyro1 form moves like the movement of objects on an old movie projector. A movie projector running at 50 fixed frames per second produces movement that "appears" to be continuous. The fixed pictures are called "moving" pictures even though they never move. This same phenomenon is true for our universe. A body of mass "appears" and in reality does move quite continuously, but nevertheless its gyro1s in their mass state are always at rest in ether space and move discretely as waves.

A second thing that "appears" to be true is that the 4-space of the universe appears to be empty. In a cave where there is no light, space appears to be just black space with no depth. It takes light to make it appear three dimensional. In a light wave, mass is moved in discrete steps like all mass is moved through ether space. Light in its maximum mass state is a gyro2 which is repeatedly at rest in ether space and is moved only as a wave. The mass state of people is always at rest in ether space. This makes it possible for people to see light that is in its gyro2 form. On the other hand, illuminated light is moved through 4-space by the creative-destructive activity forming photons. The photon is an at-rest, shaped, quantized, empty hole in ether space with the shape being two separate, side-by-side, identical right circular cylinders with one sub-hole existing in each of the two parts of ether space. The photon is formed when the gyro2 is unfolded and then filled when the gyro2 is formed in its new position. There is a linear set of these discrete holes in 4-space leading from the observed object to the eye. Whenever there are continuous trains of light waves following and surrounding one another, there are a very huge number of holes in existence at the same time. This honeycomb of four dimensional holes between an observed object and the eye truly makes ether space appear to be empty because these holes riddle ether space making it transparent. Further, the length of the strings in the string points of the universe are so extremely short that they cannot be detected by the eye, and for this reason, the four dimensional space of the universe appears to be only three dimensional, but the fourth dimension does give it bloom.

Gravity and time

The nucleus of an atom with rest mass *m* is formed from quarks that in turn are made up of gyro1s. The associated anti-mass contained in the circuits in the nucleus possesses an effective kinetic energy equal to *m c*-squared. (This is the value obtained using a local rod and time as given by a local clock that is at rest in a uniformly dense set of pocketons wherein the density gradient is zero. In this situation all gyro1s are closed and the vibrations of any one of these gyro1s could be used for a clock. Note though, that at rest, different sized gyro1s vibrate at different rates.) When the rest mass *m* is accelerated to a velocity 2*v* for half of time, the sum of the kinetic energy contained in this mass, *m*, plus the kinetic energy contained in its associated anti-mass, *m*, is shown in Chapter 10, theorem 16 to maintain the value *m c*-squared as long as one uses the same local at-rest rod and clock. Thus, at this point, kinetic energy is conserved and gravity has accelerated a molecular object without a need to produce a change in total energy.

But in order to make its way through space at an average velocity *v*, the mass has to be increased to the value *m* multiplied by the Lorentz factor (i.e. one over the Lorentz radical) which is greater than one. With this additional mass there is an associated additional amount of anti-mass, and in every respect these additions are treated the same as the original mass and anti-mass that made up the at-rest nucleus with one exception. The one exception is that the additional pocketons are unfolded from the ether space on the front side of gyro1s and are refolded back into the ether space at the same spot on the back side of gyro1s within the moving nucleus, and thus in the direction of the velocity vector their kinetic energy is equal to zero. The increase in mass, going from *m* to *m* multiplied by the Lorentz factor, increases the mass kinetic energy from m v-squared to *m v*-squared multiplied by the Lorentz factor. This increase in mass kinetic energy needs to be accounted for.

A gyro1's folding chamber unfolds or folds nuggetrons and pocketons at a constant rate that depends upon the at-rest mass of the gyro1. Let *t* be the time for one folding-unfolding vibration for a gyro1 when it is closed. Then from Chapter 10 the time for one vibration, when moving at an average velocity *v*, is *t* multiplied by the Lorentz factor. The effect of employing the creative-destructive activities over the increased time for each vibration produces an apparent mass kinetic energy for this nucleus that is equal to *m v*-squared multiplied by the Lorentz factor, and it does this with no increase in total energy. This is in complete agreement with the above kinetic energy that exists after the mass is increased. The increase in time required for each vibration of a gyro1 produces all of the required increase in effective mass kinetic energy, and as stated above, the creative-destructive activity accomplished this in an energy conserving manner. Thus, gravity produces accelerations of molecular objects without producing any change in total energy. In short, for a given mass object its folding-unfolding chambers using the creative-destructive activity always produces kinetic energy in an energy conserving manner whether the object is moving or is stationary in ether space. The mathematical details of the conservation of energy and the Lorentz transformations are explained in Chapter 10, theorem 16.

Dialogue

The following dialogue is introduced to illustrate the way gravity imparts additional speed to mass objects without making any change in the total energy of the universe.

Doctor Gravity: Mr. Rock, would you like to be made into a beautiful shooting star?

Patient Rock: I am just sitting here at rest in ether space. I am not going anywhere or doing anything. So yes, I would like to become a beautiful shooting star. Please, tell me how you will operate on me to accomplish this change.

Doctor Gravity: You have within you a large amount of kinetic energy in the circuit pocketons that your vibrating heart is sending round and round in the two circuits in each and every one of your quarks. I will need to operate on you, and convert some of this kinetic energy into kinetic energy for you personally, for by law I cannot create any new energy. Thus, the only way I can operate on you and give you, yourself, some kinetic energy is to make use of your internal kinetic energy that is already in existence. This is a simple procedure and is accomplished by merely opening all your gyro1s. However, there is a slight problem. I cannot open all your gyro1s without also, increasing your mass. The more kinetic energy I give to you when I increase your velocity, the more I must increase your mass.

Patient Rock: This worries me a little. I think I am large enough already, and I wonder, if this added mass might have some side affects.

Doctor Gravity: If I impart too much mass and kinetic energy to you, you will tend to break apart, but this still will not keep you from becoming a beautiful shooting star. In fact the added mass will make you into an even brighter shooting star. But in order to not break the law, I have to do one more very important thing. I will have to slow the rate at which your heart vibrates in each and every one of its folding-unfolding chambers.

Patient Rock: Now I am really getting nervous. Sitting here at rest in ether space, my heart always has vibrated at various rates within its various chambers, but each of these rates is invariant. If you slow my heart's rates of vibrating, I may become faint or possibly even disappear.

Doctor Gravity: Your location in space is not very far from earth. So for the velocity I will impart to you, you should not notice any change. In fact, I will slow the rate at which your watch ticks by the same factor that I slow down the rate of your heart. This means that you will not notice or observe any change in each of the rates at which your heart vibrates.

Patient Rock: Now I feel a little better. But if you increase my mass aren't you breaking the law? The reason being that when you increase my mass, you will increase my kinetic energy, which by law you are not allowed to do.

Doctor Gravity: This is a very good observation, but you are overlooking the way your heart vibrates. Your heart vibrates because the creative-destructive activity is continually functioning in each of two, folding-unfolding chambers in each gyrol that forms you. In these vibrations this activity does not change the total energy, and yet keeps you in existence by keeping your gyrols coming and going in two, out-of-sync sets that are associated with each of your quarks. Each folding-unfolding chamber creates or destroys nuggetrons and pocketons at a constant rate, but this rate may vary from disjoint quark to disjoint quark. When all the chambers of your heart are similarly operated on, and made to involve additional pocketons and nuggetrons in each vibration, your mass increases by a certain factor, and the time required for each of your hearts vibrations increases by this same factor. This means your unit of time has increased, and your watch has slowed down. At this point, the creative-destructive activity is moving both you and your additional mass in exactly the same way, and yet, like always, it is conserving total energy. Taking some of your internal kinetic energy and simultaneously slowing down your heart's vibration rate keeps me within the law when I increase your kinetic energy.

Patient Rock: It sounds to me like you know what you are doing. I am ready to begin the operation.

Doctor Gravity: I must inform you that Doctor Friction will join me in this operation when you are nearing the earth. He will use the earth's atmosphere to begin warming you, and in a short time will make you into a beautiful shooting star. I think everything is in order, and we can begin.

Patient Rock: I have complete confidence in you, Doctor Gravity. I am convinced that you will stay within the law, and I will become a beautiful shooting star.

Gravity and opening gyro1s

The following is an extended, but still brief outline of the way gravity conserves total energy. The linear unfolding rate associated with any gyro1 is always invariant when measured by clocks at rest in zero gravity. This means that the linear unfolding rate in the circuits of a quark is a maximum when the associated gyro1s are closed because then the linear unfolding is parallel to the tangent line to the circuits. When the gyro1s which form a nucleus are open to move, the unfolding lines are along the axis of the electron and along the axis of the positron in the sea of ether pocketons and now they are at an angle to the tangent line to the circuits of moving pocketons. This makes each circuit's linear rate of unfolding slow down as the velocity of the gyro1s is increased. In Chapter 10, using the fact that sine squared plus cosine squared is equal to one, it is established that at this point the increase in velocity has not caused any change in the mass kinetic energy plus anti-mass kinetic energy of these gyro1s.

But as stated above, in order to move through ether space the mass of the nucleus must be increased. With the additional mass and associated anti-mass comes additional kinetic energy and this somehow must be accounted for. The additional anti-mass pocketons are moved like the circuit pocketons and carry a cosine-type kinetic energy. Thus, again as illustrated above, the way all the additional kinetic energy is accounted for is by increasing the time required for each vibration. This increase in the time for each vibration forces the gyro1 to move more mass, but move it with a slowed down rate of vibrating. This is accomplished by merely having the fixed-rate, creative-destructive activity go a little longer on each vibration. The increase in mass creates new kinetic energy which is counterbalanced by the slowing down of time. In the end there is no change in the total energy. The fact, that this is true for all gyro1s implies it is true for all quarks. This also means it is true for any molecular body of mass which is moving as a de Broglie wave. The creative-destructive activity accelerates molecular objects in a gravitational field without requiring any change in the total energy.

The force of gravity is produced by the creative-destructive activity taking place in gyro1s that are open. The creative-destructive activity's pull on nuggetrons is strongest in the direction of the local ether pocketon's density gradient, and this causes the gyro1s to open in this direction, and consequently to produce movement in this direction. This is gravity in action. The creative-destructive activity does all of the work, and the opening gyro1s determine the acceleration and velocity as dictated to them by the magnitude and direction of the density gradients in the gravitational field that exists in the local sea of ether pocketons.

When the direction of the density gradients of ether pocketons is changed, then so is the direction of the motion. Under ordinary acceleration of mass this is the event that occurs. Also, the local density gradient field can be modified using magnets to move pocketons, and can even cause the force of gravity to be overcome. This event takes place in a levitron OMEGA. If one could discover how to control the local density gradient field of ether pocketons, objects could be made to fall in a chosen direction. Or if by some method one could get the gyro1s that make up a neutral mass object to all open in a predetermined direction, then the object would move in this predetermined direction. Shockwaves that are produced in the sea of ether pocketons by bounces of properly colliding pistons may be capable of accomplishing this feat. (See the increase in momentum as displayed in the bouncing balls experiment in Chapter 12.)

Proper theoretical approach

Given a structure that is to be studied, what is the first thing one must do? The first thing one must do is to define this structure using a complete and consistent axiomatic system. If an independent axiom is left out of the defining axiomatic system, then the structure being studied is not the desired structure, but rather one is studying a different structure.

The derivation of any theory concerning a given structure requires that, at all times, all defining axioms must be adhered to. As soon as one defining axiom is overlooked and violated in any calculation, then the obtained results probably do not apply to the structure being studied. Sometimes it is very tempting to introduce a new axiom in order to remove a confronted problem, and then to study this new axiom and completely disregard all the other known axioms or properties possessed by this structure. This is not a proper mathematical approach. An example is given below.

An improper definition of our universe

The defining axiomatic system for our universe must contain all of the universe's independent properties and must be consistent. Let it be assumed that the definition of our universe contains among other laws, the law of action and reaction, and also, a second law that states that space is empty. Known particles and anti-particles obey the action and reaction law, but a gravity that obeys this same law in empty space is impossible. Gravity produces a force on objects, but where is the local reaction force absorbed? In empty space there is nothing for gravity to push on or pull on. In empty space there is no traction. Thus, there is something wrong in a definition of our universe which contains both of these stated laws.

A housing Euclidean space

A housing Euclidean space is required for many calculations in our universe. In order to use a housing Euclidean space for calculations, one needs to establish a unit of time, a unit of length, and a unit of weight. This brings up the question, "Do people have the correct impression of their height, weight, and how fast time is going?"

It may be that adult people are really as tall as they envision elephants as being, and the elephants are really much taller. Or it may be that adult people are really as tall as they envision rabbits and rabbits are much shorter. There really is no way of determining a true, absolute measure of length. All one can do is choose a unit of length and compare all other lengths to this unit of length.

It may be that adult people are really as heavy as they envision an elephant as being, and elephants are really much heavier. Or it may be that adult people are really as heavy as they envision rabbits and rabbits are much lighter. There really is no way of determining true, absolute weight. All one can do is choose a unit of weight and compare all other weights to this unit of weight. (It is true that the number of nuggetrons that are used to form a given object determines its mass, and each nuggetron is composed of half a Planck's constant of energy. But in order to state Planck's constant, established units are required.)

It may be that what people envision an hour to be is really a day of time and a day is really much longer. Or it may be that what people envision an hour to be is really more like a minute and a minute is really much less time. There really is no way of determining true, absolute time. All one can do is choose a unit of time in a given setting and compare all other unites of times to this unit of time.

Picked units of length, weight and time are used in a Euclidean space. One possible set of units is the centimeter, gram, second system. One may feel comfortable doing calculations in Euclidean space using these units. But the units in this chosen set, or any other chosen set, are highly variable. When these units of measure are moved about in the universe of unitivity theory and viewed from a fixed housing Euclidean space, because of the change in density of the ether pocketons and associated nuggetrons they are rarely going to agree with their original chosen values. It appears that this fact should make it difficult to do calculations in the space of our universe, but this is not true. All nuclei are made from gyro1s. This makes all molecular things change in size and time rate together. This, along with the fact that gyro1s satisfy the Lorentz transformations, means that most problems are avoided, but still these phenomena should be studied further.

Outside-in view added to inside-out view

When individuals study the universe this amounts to the universe studying itself, because all individuals are part of the universe. This is something like an eye trying to look at itself and determine its color. The eye is good for looking from the inside-out, but is not good for looking from the outside-in and trying to see itself. The eye needs a mirror to help it look at itself. Similarly, for individuals to study our universe they need a magic mirror to help them look at our universe from the outside-in. This book illustrates that this required magic mirror is mathematics.

For discussion:

1. How reliable are theoretical results?
2. Given a real structure, how can one prove that it has been defined properly?
3. What is mathematics?
4. In what way is mathematics a magic mirror for viewing our universe?
5. How absolute are units of time, length and weight?
6. How is energy conserved in a mass object in a gravitational field, when the object is being held and not allowed to fall? When gyro1s open the kinetic energy of the anti-mass in the internal circuits is decreased. Where does this kinetic energy go? (Hint: Consider how the unfolding and folding movement of pocketons influences the local density gradient and the local nudging process.)

4 Unitivity Theory: A Theory of Everything

Introduction

This chapter gives a brief, but quite complete, derivation of unitivity theory. Unitivity theory as presented in this chapter is based on a material axiomatic system, and the beginning part of this chapter is similar to Chapter 2. The stated material axioms disclose that our universe consists of a three-dimensional empty space that exists between two, thin, four-dimensional ether subspaces. These axioms along with observations reveal properties of mass, anti-mass, energy, gravity, electricity, magnetism, momentum and light; and provide information about relativity theory.

One unique feature of unitivity theory is its strong emphasis on the conservation of energy. Von Neumann's insistence on conserving energy in his calculations in the field of hydrodynamics, lead him to many successful results. Working at the Los Alamos Scientific Laboratory and becoming familiar with some of his work convinced the author of this book that this law should never be violated. If you study unitivity theory, you will find that the universe has many elegant and clever ways of conserving energy.

Unitivity theory gives a structure for the universe that is rich enough to carry a theory of everything, i.e. to carry a TOE. A request is made to the scientific community to take this theory into the laboratory and check it out. Wherever it checks, use it, and if it does happen to fail in some area, modify it. Studying unitivity will help one avoid using a structure for the universe that is so thin that it is not capable of carrying a TOE, and consequently, it will help one avoid trying to find a TOE in a structure where a TOE does not exist. Studying unitivity theory helps one avoid attempting to do the impossible.

The universe defined

In order to understand the universe one must determine all of the independent properties, (or laws, or axioms which ever one prefers to use) that define our universe. No independent property can be derived mathematically from other independent properties. If one can obtain a new property from a given set of independent properties, then this new property is not an independent property, but rather it is a dependent property.

Material axiomatic system

For unitivity theory the independent laws that are used to define the universe are based on observations. The material axioms used are the following:

1. For every action there is an equal and opposite reaction or anti-action in the sense that if the two are combined, there is no action.

2. For every change in energy there is an equal and opposite change in anti-energy in the sense that if the two are combined, there is no change in total energy.

3. Two independent energies cannot coexist in the universe at the same point at the same time.

4. There presently exists a creative-destructive activity that repeatedly forces basic mass particles into existence in quantized circular loops out of empty space, and alternately operates in reverse and forces each and every one of the created basic mass particles back out of existence in such a way that, if at any time all the energy and anti-energy are combined, they completely cancel one another.

In addition to these material axioms there is a fifth axiom that is taken to be true. This axiom is needed in order to make it possible to obtain a unique structure for the universe instead of a huge number of possible structures. The needed axiom is the following:

5. The structure of the universe is as elementary as possible.

This axiom implies that, whenever there is already an available object in existence for completing a task, it must be used instead of introducing something completely unrelated and new into the structure of the universe.

Contributions of observations

In addition to accepting these axioms as true, unitivity incorporates observed truths of modern physics. One example of these truths is that the energy associated with light is quantized. This indicates that the smallest possible amount of energy possessing a neutral charge is given by Planck's constant when one is working in the cgs system. For the equal amounts of quantized positive and negative energy of unitivity theory, it follows that their smallest possible amount of energy is half of Planck's constant. These smallest possible positive and negative amounts of energy are called positive and negative nuggetrons, respectively. Let us now proceed to use the above laws to determine the nature of these smallest basic particles.

Smallest Basic particles

Using axiom 4 to produce a single nuggetron will violate axiom 1 in that there is an action, but no equal and opposite reaction. Thus, for each, say positive nuggetron produced there must be another identical, but made backwards, nuggetron. This second type of nuggetrons is designated a negative nuggetron. Thus, positive and negative nuggetrons must come paired.

Quantized loops

The most elementary way to obtain a larger positive particle is to form a clockwise oriented, quantized loop (doughnut) of positive nuggetrons provided a symmetric counter-clockwise, quantized loop (doughnut) of negative nuggetrons is formed simultaneously. (It is theoretically possible that there are also, other matched configurations.) These quantized loops must originate in pairs and need to appear as coexisting in three-space in order for axiom 1 to be satisfied.

These quantized loops have a "matter antimatter" relation which is not to be confused with the "mass anti-mass" relation discussed below.

The matter antimatter relation is associated with the action and reaction axiom while the mass anti-mass relation is associated with both the conservation of energy axiom and the action and anti-action axiom.

The matter antimatter relation is mainly positive nuggetrons and offsetting, negative nuggetrons, along with clockwise and offsetting, counter-clockwise motion.

The mass, anti-mass relation is a positive nuggetron and the soon to be introduced positive pocketon relation, or a negative nuggetron and the soon to be introduced negative pocketon relation. This mass anti-mass relation involves energy and anti-energy along with action and anti-action. If one views a nuggetron as a source of energy, then its associated pocketon is a sink of energy, and when the two are combined, there is no energy. When the action on a nuggetron is combined with the anti-action on its associated pocketon there is no action.

Relation of mass to charge

It is interesting to note how mass and charge are related. For any increase in the mass of an entirely positive mass object, say an increase in mass of n positive nuggetrons, then there is at the same time a proportionate increase of n basic units of positive charge on the object. The creative-destructive activity produces the properties of both mass and charge during the unfolding and folding. This means the greater the mass of an entirely positive or entirely negative charged particle, the greater is its charge.

This shows that adding one more positive nuggetron to a given positive loop does not change the ratio of mass to charge. However, not just any number of positive nuggetrons can be added into a quantized positive loop. Rather, there are commensurate conditions on the size of loops which must be met. (Positrons cannot get large without bound.)

Similar statements hold for negative nuggetrons and negative quantized loops. It is this relation between mass and charge that makes it possible to have the different colors of light. (See Chapter 10.)

Positive energy

Quantized loops agree with observations and have names such as positrons and electrons. By observation, it is known that both positive and negative mass are positive energy, because they both contribute to the energy of an atomic bomb. Also, as will be established, they both contribute energy equally in the equation $E = m$ times c-squared. Pushing in opposite directions can conserve action and reaction, but both of the objects being pushed and accelerated will have an increase in kinetic energy. This means that axiom number two is not satisfied when only two smallest basic types of particles are produced.

Conservation of energy

What is needed is another type of energy that can be combined with the energy of these nuggetrons making it possible to return to no energy. This new type of energy is called anti-energy or anti-mass. (One may prefer the term dark-energy.) This requirement implies that two new basic particles must be introduced. One that when combined with a positive nuggetron takes it back to zero energy, and a second that when combined with a negative nuggetron takes it back to zero energy. These particles are called positive pocketons and negative pocketons, respectively. In order to preserve the action and anti-action required by axiom one, these two new types of particles must be incorporated with the two types of mass particles so that in the end all actions are offset with corresponding reactions or anti-actions. If the nuggetrons are forced into existence and have positive energy, then the pocketons must be forced into existence in an opposite action sense and be made oppositely in an energy sense. That is, there is an anti-force associated with the forming of pocketons of each type that is offset by the force on forming nuggetrons of their related type, and in addition this anti-force produces anti-energy instead of energy.

The two symmetric subparts of space

Further, and this is very important, for axiom 1 to hold, each of the created basic particles must be placed into a space which is symmetrically equivalent to the space into which its associated, backwardly-made particle is placed. In addition, both types of mass particles must be placed away from their respective anti-mass particles in order to make it possible to control the activity of their coming and going.

This can be done, if and only if there are at least two adjacent and symmetrically equivalent subspaces that together makeup all of the universe's space. The law of action and reaction demand that these two, adjacent, symmetric sub-worlds exist.

Four-dimensional universe

In unitivity theory these two subspaces are described in a very elementary way by using a four-dimensional space. Each pocketon's region of influence is taken to be a four-dimensional hypercube. The positive pocketons are placed one layer deep in the positive direction of a given orthogonal base vector for the fourth dimension. To form the second part of space adjacent to this first part, the negative pocketons are placed one layer deep in the negative direction of this same given orthogonal base vector for the fourth dimension. When the lengths in the direction of the fourth dimension are all set equal to zero, these two subparts of ether space appear to coexist as a single, empty, three-dimensional space. This is a very interesting mapping, in that when the very "short" fourth dimensional lengths which cannot be seen are set equal to zero, then the limited four-dimensional universe with its two ether subparts becomes a three-dimensional universe possessing only one part which is empty.

String points

This structure for ether space is best understood by noting that each point in the three-space of the universe is a string point. These string points consist of the given point and two, independent, "short" open strings. One of these strings is a short line segment in the direction of a positive fourth-dimension orthogonal base vector, and the other is a short line segment in the negative direction. The length of any string is the length of a side of the four-dimensional hyper-cube that exits at the given point. This not only means that ether space can be warped, but in addition each of the two subparts of ether space can be warped basically independently. To really visualize these concepts, it is helpful to study the two dimensional universe as presented in Chapter 2 of this book.

The activity of axiom four

The activity that brought the universe into existence out of empty space has two phases. In the folding phase, a positive nuggetron and a positive pocketon, and a negative nuggetron and a negative pocketon are created as a foursome with the nuggetrons having opposite charges and with each nuggetron and its associated pocketon being pushed apart. In the unfolding phase the mass and the anti-mass are destroyed as each nuggetron and its associated pocketon are pulled back together and out of existence. Note that in order to produce two parallel subparts in space, the basic particles must come as foursomes, but once in existence they can be destroyed as twosomes, and yet have both the conservation of energy and the law of action and anti-action satisfied.

Basic structure of the universe

The universe is now simple as one, two, four. There is one activity of mass interacting with anti-mass. There are two symmetric and side-by-side subparts to ether space. There are four most-basic types of particles. This is "the simple as one, two, four universe" of unitivity theory. Note that it is not possible for the universe to be as simple as one, two, three.

Properties of particles

What properties must mass and anti-mass particles possess? Mass and anti-mass possess a canceling effect, thus their properties must be canceling or opposite in nature. Mass is known to be concentrated in ether space, thus to be opposite, anti-mass must tend to expand and fill ether space. When mass particles are squashed together, one encounters a resistant, pushing out-ward force. When anti-mass particles are squashed together, in order to be opposite in nature, one must encounter a nonresistant, pulling in-ward force. For mass, the reaction to squashing is push-out and for anti-mass the reaction to squashing is pull-in.

Mass particles are quantized and shaped by the "presence" of nuggetrons. On the other hand space particles (or force-carrying particles) are quantized and shaped in an opposite way by being the "absence" of pocketons. This is to say that space particles are just quantized shaped empty space. For this reason space particles (or force-carrying particles) may be viewed as vacuum energy particles in that they consist of a true vacuum in the sea of ether pocketons. Using other words, force-carrying particles are shaped, quantized, vacuum energy in a sea of dark-energy.

Density properties of space

Let us consider the change in the density of pocketons in the presence of a change in the number of nuggetrons which are acting as a source-point mass. Law 3 implies that added positive nuggetrons in the presence of negative pocketons in their associated subpart of ether space, and/or added negative nuggetrons in the presence of positive pocketons in the other subpart of ether space, are going to cause squashing in that none of these can coexist in four-space. Consequently, the squashed ether pocketons of each type pull themselves together and increase their density as well as the density of their respective, encased nuggetrons.

Gravity

It is well known that the force of gravity is an increasing function of the total amount of neutral mass that is acting as the field's source-point mass. That is, an increase in the amount of neutral mass proportionately increases the force of gravity. This is why the acceleration of gravity is greater on the earth than on the moon. The pertinent question is, "How is this formula for the force of gravity arrived at in unitivity theory?"

The activity of either simultaneously creating, or simultaneously destroying mass along with its respective anti-mass is the only activity that exists in unitivity theory. Thus, this is the only activity available to produce the forces associated with gravity, electricity and magnetism.

Directed lines of force

To explain why the force on each unit of neutral mass is greater on a falling object on the earth than on the moon, unitivity requires and uses pocketon density. For this discussion, the introduction of directed-lines of pull-force between pocketons in 3-space, and directed-lines of force upon nuggetrons also, in 3-space, is advantageous.

The three dimensional directed-lines of pull-force on the ether pocketons, for each type of pocketons, are directed in the direction of the respective ether pocketons' density gradient. (That is, in the direction of maximum increase in density of that type of ether pocketon.) The number of directed-lines of pull-force per square unit between adjacent pocketons is defined to be proportional to the magnitude of the pull-force per square unit that is produced when there is a density-difference between adjacent pocketons. The greater is the density-difference, the greater is the number of directed-lines. (The magnitude of the pull-force per square unit in 3-space that pulls two adjacent pocketons together is very large due to the extremely large family of 3-dimensional surface areas these adjacent pocketons share in 4-space.)

In unitivity theory an increase in the density of each type of ether pocketon causes the activity of mass interacting with its associated anti-mass to have a proportionate increase in the pushing-force (during the folding phase), and the pulling-force (during the unfolding phase) upon each of the involved nuggetrons, and to have a proportionate increase in the offsetting anti-force acting upon each of the involved pocketons. These three dimensional pushing and pulling forces upon each of the involved nuggetrons are equal in magnitude, and they are both in the same direction which tends to be in the direction of the local directed-lines of pull-force between the ether pocketons. For this reason one can use just the number of directed-lines of force to represent the magnitude of the pushing or pulling force upon each nuggetron. Putting this together, locally the change in the number of directed-lines of force acting upon each nuggetron, as produced by the creative-destructive activity, is proportional to the local change in density of ether pocketons. Consequently, it is also proportional to the change in the number of directed-lines of pull-force per square unit acting between the local ether pocketons.

The greater is the amount of source-point mass, the greater is the squashing of the local pocketons; which implies the greater is the number of directed-lines of pull-force per square unit between the local ether pocketons; which implies the greater is the change in density of the local ether pocketons; which implies the greater is the change in the number of directed-lines of force acting upon the local nuggetrons.

Gravitational formula

Checking this out on a three-dimensional sphere, whose center is the given source-point for the neutral mass that produces the gravitational field, yields the following: The rate of increase in surface area of a sphere with increasing radius, *r*, is proportional to *r*-squared. Conserving the directed-lines of pull-force on the involved ether pocketons, one determines that the number of directed-lines of pull-force per square unit of surface area that are acting upon adjacent ether pocketons drops off like one over r-squared. From above, this means that both the density of pocketons and the number of directed-lines of force acting upon nuggetrons drop off like one over *r*-squared. Thus, one obtains, "the number of directed lines of force acting upon a nuggetron, of either charge within a given gravitational field, is proportional to the amount of neutral mass acting as the source-point mass for the field, and inversely proportional to *r*-squared where *r* is the distance from the nuggetron under discussion to the gravitational source-point." This is unitivity theory's derivation of the formula for the force of gravity on mass.

Note that a neutral mass body obeys the formula for the force of gravity by equally involving the two identical subparts of ether space, but in the above derivation it is seen that this formula also applies to any charged nuggetron alone. A gravitational field is produced by nuggetrons acting as neutral mass. This explains why the formula for the force of gravity normally has neutral masses m and m' in its numerator. An electric field is produced by nuggetrons acting as charged particles, thus, the formula for the electrical force on charged particle normally has charges q and q' in its numerator. The nuggetrons produce a field the same way whether they are looked at as being mass or charge. Thus, both formulae have an r-squared in their denominators indicating the universe that we observe truly is three dimensional. (Recall that the mass of a nuggetron and its charge are proportional. Thus, mass and charge can be interchanged provided the involved units are handled properly.)

Density of pocketons calculations

The reason a three-dimensional calculation can be made for density of pocketons rather than a four-dimensional calculation, is that when making a four-dimensional, density calculation in a region of constant density of pocketons, the length of the required fourth-dimensional variables in the numerator and in the denominator are equal and cancel out. This is true because each of the two subparts of ether space is only one pocketon deep in either the positive direction or the negative direction of the orthogonal, fourth-dimensional, base vector. This means the computing of the density of ether pocketons is a three-dimensional calculation in a housing Euclidean space.

Unified field theory

The fact that there are two, related, adjacent, independent seas of ether pocketons, (or if you prefer, the fact that there are two, adjacent, independent subspaces that are filled with two different types of dark-energy) provides the universe with a rich enough structure for it to possess a unified field theory.

There are now two ways to cause a squashing of pocketons. One way is to place additional nuggetrons in a neighborhood of a given point, and the other way is to place additional pocketons near a given point.

When additional negative nuggetrons are placed in the neighborhood of a given point in the sea of positive pocketons, and additional positive nuggetrons are placed in the neighborhood of some other "near-by" given point in the sea of negative pocketons, a set of directed-lines of pull-force is formed in each of the two subparts of ether space. These are the directed lines of pull-force needed to produce the directed-lines of force on charged nuggetrons that is called an electric field. As always, in a three dimensional, spherical setting the directed lines of pull-force produce a density that drops off like one over *r*-squared in each of the subparts of ether space. Nuggetrons carry both charge and mass. In an electric field one tends to view the involved positive and negative nuggetrons as charged particles instead of as mass particles, but no matter how one views them, they squash the ether pocketons the same amount.

Neutral gyro1s can only open symmetrically. The reason being, the two circuits of rotating, strongly-held-together pocketons have to be treated equally in order to maintain the existence of the associated quark. In an electric field the positive quantized loop and the negative quantized loop of a gyro1 cannot be treated symmetrically. Thus, the movements of gyro1s are basically unaffected by an electric field.

If the same number of positive nuggetrons and negative nuggetrons are introduced near one and the same given three-dimensional point, then the directed-lines of pull-force and therefore, the directed lines of force are identical in the two subparts of ether space, and this is called a gravitational field. It is established below that in a gravitational field the neutral-mass gyro1s open symmetrically for movement, and consequently one tends to view the involved nuggetrons that produce a gravitational field as mass particles and not as charged particles. But again no matter how one views the nuggetrons they squash pocketons the same amount.

When additional positive pocketons are introduced near a given three dimensional point in the sea of positive pocketons, the squashing produces an increase in the density that is similar to that was produced when additional negative nuggetrons were introduced near a given point in the positive pocketons. A similar statement holds for the negative pocketons and positive nuggetrons. The additions of anti-mass pocketons into ether space at two different points, one point in each of the two subparts of ether space, produce directed-lines of pull-force on the ether pocketons that in turn produce the directed-lines of force on charged nuggetrons in each of the two subparts of ether space, and this field is called a magnetic field. As above in an electric field the directed lines of pull-force on the pocketons produce a density, in the respective subspaces of pocketons, that drops off like one over *r*-squared, but here there is additionally the nudging process operating.

Note that similarly to an electric field, neutral gyro1s cannot open symmetrically in a magnetic field. Consequently their movements are mainly affected by the pushes and pulls upon the pocketons which have two different densities within the magnetic field, and upon the eating up of space itself between two unlike poles. This is the way magnets are attracted to one another. In contrast however, the movements of charged particles are greatly affected by magnetic fields.

Pocketons tend to nudge themselves back into one to one equilibrium in the two parallel subparts of ether space, (They move like trains not like flowing water.) for this reason magnetic fields require the pocketons continually to be replaced in order to maintain the existence of the magnetic field. This completes the derivation of the unified field theory that unitivity theory possesses.

Gyros of the first kind

The pumps that move pocketons around are the gyros of the first kind and are designated as gyro1s. What these gyro1s are, and what some of their properties are, will now be discussed.

The gyros of the first kind consist of two paired quantized loops. One of these loops is composed of positive nuggetrons and the other is composed of the same number of negative nuggetrons. One of these loops has a clockwise orientation or spin while the other has a counter-clockwise orientation or spin. In order for these to be stable and at rest when folding and unfolding, the two quantized loops must be intermingled. That is, they must appear to coexist in three-space. In unitivity there is room for all this to take place due to the fact that there are two, thin, adjacent, four-dimensional subparts of ether space that face each other at each point in the universe's three dimensional, empty subspace.

The pocketons are moved by the folding and unfolding that takes place in these gyro1s. The linear unfolding rate along the axis of a gyro1 (the length of unfolding per unit of time as measured by a near-by clock and rod both of which are at rest in zero gravity) is always $2c$. In order to satisfy axiom one for a given gyro1 at rest, the two types of pocketons must move in opposite directions perpendicular to the common face of the coexisting three-dimensional loops. As viewed from above, a clockwise oriented quantized loop removes pocketons from below the loop and replaces them above the loop. A counter-clockwise quantized loop moves its associated pocketons the opposite direction from that of the clockwise loop. (See figure 1.) This activity, for a gyro1 at rest in the sea of ether pocketons, satisfies the law of action and reaction, and conserves energy.

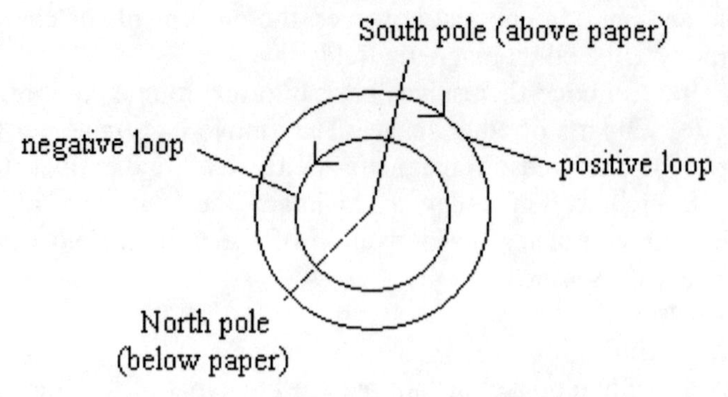

Figure 1. Gyro of the first kind

Magnets

A gyro of the first kind is a small symmetric magnet. In a gyro1 a number of positive pocketons and an equal number of negative pocketons are moved symmetrically. On the side where the positive pocketons are removed and the negative pocketons are placed, a north pole is formed, and on the side where the negative pocketons are removed and the positive pocketons are placed, a corresponding south pole is formed. At a north pole there is an excess of negative pocketons compared to the number of near-by positive pocketons. At a south pole there is an excess of positive pocketons compared to the number of near-by negative pocketons.

(It should be noted that an electron alone or a positron alone is a magnet, too, but not a symmetric magnet. For each of these particles, one pole is formed by removing pocketons of one type from ether space during the unfolding phase, and the other pole is formed by the adding of these same pocketons back into the same subspace at a different point during the folding phase. This produces a surplus of pocketons of the opposite type from those removed at the first point, and a surplus of the replaced type at the second point.)

When a north pole of one gyro1 is placed near a south pole of a second gyro1, due to the resulting change in densities of pocketons, in the gyro1 having the north-pole the pushing and pulling apart in the sea of the negative pocketons is reduced and the pushing and pulling together in the sea of positive pocketons is increased. In the gyro1 having the south-pole the pushing and pulling together in the sea of the negative pocketons is increased and the pushing and pulling apart in the sea of the positive pocketons is decreased. Further, when a north pole is close enough to a south pole, a north pole removes positive pocketons and a south pole removes negative pocketons from between them faster than the nudging process can replace them. This leaves the two poles under discussion setting there with no room left between them. This means a north pole has been attracted to a south pole, and vice versa. Also, the pushing and pulling forces produce a momentum field, and this field can aid in the movement of magnets. When the two close together gyro1s are out-of-sync in their folding and unfolding then they share pocketons. (That is the pocketons one gyro1 is folding are simultaneously being unfolded by the adjoining gyro1 and vice versa.) The end result in this case is the destruction of the two poles that had existed between the two gyro1s. This in turn makes the two gyro1s become a single, longer magnet.

Similarly, when two like poles are placed near each other, the pushing and pulling apart is increased and the pushing and pulling together is reduced, and in addition the pocketons of either one type or the other are being placed between the poles faster than the nudging process can remove them. This means that like poles repel each other.

Larger magnets are produced by involving a number of gyro1s that are "properly lined up, and are acting independently." By heredity, for all larger magnets, opposite poles attract each other and like poles repel each other.

Review

To make some of this a little clearer, some things will be restated and a few things will be added. In unitivity the form for a quantized loop is a doughnut. The form for a gyro of the first kind at rest in three-space is composed of two coexisting doughnuts with opposite spin and of opposite sign. The direction of the folding and unfolding is along their common axis which is perpendicular to the face of the two coexisting donuts. Thus, for a gyro1 at rest in ether space the two types of circuit pocketons are moved in opposite directions, perpendicularly to the common face of the coexisting, quantized loops. This means that each gyro1 is a magnet. This is all possible because the different types of circuit pocketons exist in different, but adjacent, four-dimensional subspaces.

Moving gravitational fields

Mass is encased in the sea of ether pocketons, and is always at rest with respect to adjacent ether pocketons. If the ether pocketons shift by expanding and/or contracting, then the encased mass moves along with this movement. This shifting movement can be studied by employing a housing, three-dimensional, Euclidean space. For "large" bodies of mass revolving around one another, it is easy to show that moving gravitational fields are the major source of their movements. This fact makes the constancy of the speed of light hold approximately, but not exactly, for "large" bodies of mass. The larger the body happens to be, the better is the approximation. However, this type of movement does not influence time. In unitivity the velocity that actually causes time to slow down is the velocity of a gyro1 moving through the sea of ether pocketons. The nature of this discrete movement through dark-energy will now be discussed.

Gyros of the first kind in quarks and de Broglie waves

When a gyro of the first kind is in a quark, it is tied into the quark by being paired with two, (one on either side), like gyro1s which are completely out of phase or out of sync with it. The way quarks are formed from swirls is discussed below. In a swirl a gyro1 no longer can move as an individual magnet. Rather, the two quantized loops which coexist when the gyro1 is at rest have to pivot open symmetrically in three-space in order to form a hole in each of the two seas of ether pocketons. The gyro1 is then capable of moving through the locked together ether pocketons, and is still capable of equally maintaining the two circuits of rotating pocketons. This is explained in detail in Chapter 10 where gyro1s and swirls are formally introduced.

For calculation purposes, replace each charged doughnut in any given gyro1 at rest with the smallest circle in three-space that fits around it, and orient these coexisting circles so that they possess a diameter that is parallel to the direction of motion. (The pushes and pulls of mass interacting with ant-mass in a gravitational density gradient field produce this directional orientation and simultaneously cause any involved gyro1 to open.)

Now for gyro1 movement, these coexisting circles pivot open symmetrically about a second diameter that is perpendicular to both the directional diameter described above and to the axis of rotation of the given gyro1. For the two coexisting circles there is only one such second diameter. To change directions, a second directional orientation adjustment of the two circles is required.

The greater is the density of the local ether pocketons the greater is the push and pull forces of the creative destructive activity. This implies that the amount a gyro1 pivots open in a gravitational, density gradient field is proportional to the density of the ether pocketons that encase it.

By heredity, the way gyro1s open to move quarks as waves describes the way mass moves as de Broglie waves. (See Chapter 10.)

Momentum

For a gyro1 falling in a gravitational field, its total associated density gradient vector field is the vector sum of the gravitational and momentum gradient vector fields. During the folding phase of a moving gyro1, the pushing increases the density of the forward, adjacent ether pocketons.

The greater is the total amount of moving mass, and the greater is the velocity, the greater is the increase in density. This is how folding produces the momentum gradient field. This shows that a gravitational field and a momentum field are closely related.

The movement of a gyro1 in the case where the directed-lines of force on nuggetrons are identical in the two subparts of ether space is very common. This is due to the fact that neutral mass is very common. Note that any moving, open gyro1 will require a restraining force in order to close and come to rest. When it does come to rest in ether space, its momentum field is obliterated, but any local gravitational field still remains. An at-rest gyro1 in a gravitational field remains somewhat open, and thus, has weight and also produces a nudging of ether pocketons.

Spread angle

In unitivity the spread angle is half of the angle between the planes that contain respectively, the two looped circles of a gyro1. As the gyro1 opens the spread angle is increased and the velocity of the gyro1 going through the sea of ether pocketons is increased. (See Chapter 10.) The nature of this increase in velocity will be discussed next.

Linear unfolding rate

The unfolding process for any quantized loop forms a quantized shaped hole in the sea of ether pocketons. This quantized shaped hole is simply empty space. The linear rate at which this hole is formed for each involved loop, has the constant value $2c$ when measured by a near by, at-rest rod and clock where gravity is neglected. (This c is also the average speed of all light through the sea of ether pocketons in the present form of unitivity theory. In unitivity theory as it now reads, the switching from the folding to unfolding phase is taken to be instantaneous. If the switching is not instantaneous, then blue light travels a little slower than red light. This is because blue light has a higher frequency, and as is established later, has more switches from folding to unfolding, and from unfolding to folding per second than red light. The author understands that this property of light is being checked by NASA. No matter what the result of this experiment turns out to be, the linear unfolding rate of $2c$ is not affected.)

Lorentz transformations

Using the above spread angle and trigonometry it can be proved that gyros of the first kind in quarks act like analog computers and produce the Lorentz transformation formulae for mass, time and length of a rod as functions of velocity. (See Chapter 10.) This implies that in unitivity theory, time and mass change with increasing speed, and that there is a maximum speed at which mass can move. Also, it follows that there exists a weak form for the constancy of the speed of light. It states that at the gyro1 level when an observer is moving at an apparent speed v through the ether pocketons, then $c-v$ is mapped to c, but for $v>0$, $c+v$ is "not" mapped to c.

The weak form of the constancy of the velocity of light

The weak form of the constancy of the speed of light is weak for the following reason. The rods and clocks of a moving observer will, at the gyro1 level, measure the speed of a light signal to be c when the signal is going in the same direction that this observer is moving. (It should be noted that in unitivity theory it is proved that rods shorten at the gyro1 level. It then follows that this is true at the quark level, but in unitivity theory it may not be true at the molecular level. But by heredity, the time and mass changes do hold in the "large.") This c is the same as that measured by an observer at rest measuring a light signal. But in unitivity theory, a signal coming straight toward the moving observer from in front will have a measured speed value of approximately $(c+2v^{\prime})$ where $v^{\prime}$ is the average speed of the observer going through the sea of pocketons as measured by the observer's moving clock which has slowed down and the observer's moving rod which may or may not have been shortened. The exact change in length of a "sizable" rod due to movement through ether space is left open in unitivity theory. Recall that here v is the observer's velocity as measured by clocks and rods at rest in ether space with zero gravity.

As observed in a housing Euclidean space these two speeds are relative and have the values $(c-v)$ and $(c+v)$ respectively. The standard, strong form of the constancy of the speed of light in relativity theory states that the moving observer's second measured value above also, should be c, just the same as that obtained by an observer at rest. This happens to be impossible for light waves and de Broglie waves moving in the ether space of unitivity theory.

Relative velocities of light waves and de Broglie waves

As stated above, viewing light from a housing Euclidean space reveals that light waves and de Broglie waves are relative like all other known waves. As viewed from "with-in" the universe, the light rays that are moving in the same direction as, and parallel to, the movement of an observer will at the gyro1 level be clocked to have the same speed c as that clocked by an observer at rest. Again, this is the weak form for the constancy of the speed of light which holds in unitivity theory at the quark level.

The more important constancy of the speed of light that holds in unitivity theory is the following: If any two linear chains of n pocketons, any where in the universe, simultaneously start unfolding by linearly combining with n nuggetron, then the two will finish simultaneously. This is as viewed from a housing Euclidean space. This means that the amount of time this unfolding event requires, as measured by a near-by, defined clock and rod, both at rest in ether space with zero gravity, is always the same at any place in the universe. This linear unfolding velocity is $2c$.

Force-carrying particles

A neutral hole, (i.e. a hole that has identical sub-holes in the two subparts of ether space) in the sea of ether pocketons is formed when a gyro1's quantized loops are symmetrically spread open and unfolded in movement. This hole or vacuum is a neutral force-carrying particle. It is characterized by a shape and a quantized absence of ether pocketons. If the opening of a gyro1 is caused by a gravitational field, the force-carrying particle produced is called a graviton. If the opening of a gyro1 is caused by the momentum field that is produced by a pushing force on a gyro1, then the force-carrying particle is called a neutral boson. This again shows that gravity and momentum are very closely related.

Swirls and quarks

Gyros of the first kind are little magnets. This implies that two gyro1s are attracted together with a north pole of one attracted to a south pole of the other. If these two adjacent gyro1s happen to be completely out of phase (I.e. out of sync, in that one is folding while the other is unfolding and vice versa.) then the two gyro1s become a longer magnet. Dense pocketons pull together harder than less dense, consequently a little bend in a line of these gyro1s will cause the gyro1s to form a distorted circle. There are two circuits of pocketons in this circular arrangement that is called a swirl. The positive pocketons are going clockwise and the negative pocketons are going counter-clockwise or vice versa. The best description of this activity for each gyro1 is "a wheel of pocketons rotating in opposite directions in the middle of a wheel of nuggetrons rotating in opposite directions where the axes of these two wheels are perpendicular to each other." Or one might say that this activity produces "a wheel in the middle of a wheel."

A force-carrying particle within each of these circuits of a given, at-rest swirl is shaped and quantized, but is virtual in that it never is completely formed. These vacuum-energy holes are being formed and filled simultaneously. The two, quantized loops of the involved gyro1s are pushed together in the folding stage and pulled together in the unfolding stage. These virtual force-carrying particles, existing in either the positive pocketons' or negative pocketons' circuit, are called positive and negative gluons, respectively. Due to the symmetry of the two circuits, two associated charged gluons always come paired, and when they are considered to be together, they form a neutral gluon. The two circuits of strongly held together circuit pocketons along with their respective gluons produce the strong force that holds quarks together. These same "circuit" pocketons when existing as "ether" pocketons make it possible for gravity to bring galaxies together. This indicates how really strong the force is that holds quarks together.

Unstable swirls

A single circular arrangement of out-of-sync gyro1s (or a swirl) is unstable, because as viewed in three-space, one of these circuits of moving pocketons will be on the inside of the common circle of rotation and the other on the outside. The reason this is true is that the two types of charged folding-unfolding chambers cannot coexist, but rather they can only operate when they are separate from each other in 3-space. Thus, the two circuits are not of the same length and therefore, are not quite commensurate. To make this swirl arrangement stable, the two circuits of pocketons in this swirl must be blended together with the circuits of one or more different swirls. An inside or short circuit of pocketons must be matched with at least one longer outside circuit of pocketons of the same type. Similarly, an outside circuit must be matched with at least one shorter inside circuit. According to unitivity theory this circular swirl standing alone is the recently discovered, unstable para-quark. Two or three circular swirls that are properly matched can form two or three quarks respectively. For a stable particle made of these gyro1s, the combined circuits for one type of pocketons must be "nearly" commensurate (be composed of the same number of pocketons and have "nearly" the same length) as the combined circuits for the opposite type of pocketons. The closer to being perfectly commensurate, the longer is the half life of the associated particle.

Atoms

The nucleus of any atom is made from a number of combined matched quarks. This means that the nucleus of an atom is made from sets of gyro1s completely out of sync. Thus, for the nucleus of any atom at rest the amount of mass that exists in the mass state at any time is invariant. Half of the total mass is in the wave state and half is in the mass state. Also, in the nucleus of any given atom at rest, the rate at which the mass is unfolding and folding is invariant. The way gyro1s open in a gravitational density gradient field causes a component of the folding and unfolding in any given swirl to be perpendicular to the plane that contains the face of the swirl. This produces movement or weight in the direction of the directed-lines of pull-force of the ether pocketons. This shows that the weight of any atom at rest in any isolated and constant gravitational field is invariant. Further, the sharing of circuit pocketons inside quarks, frees approximately half of the pocketons associated with atoms to be used for filling the two subparts of ether space with pocketons. (The fact that the light radiated from atoms has colors that are uniquely determined by each different atom, proves the existence of matched internal circuits. The orbiting electron and a gyro2 plus some other things come from these circuits when a proton is produced from a neutron. The positron that is associated with the orbiting electron remains in the positive circuit within a newly formed proton giving the proton its charge and making it extremely stable. Seeing how hard it is to match up all the circuits needed to form an atom, it is not surprising that only 4% of the energy of the universe is found in atoms. See Chapter 11 for the complete breakdown of energy of atoms, energy of dark matter, and dark energy, and how this observed breakdown verifies unitivity theory.)

Gyros of the second kind

The second way that two matched quantized loops (doughnuts) can form a stable structure in three-space is for them to have opposite spin orientation and be tangent to each other on a common parallel plane. This is in contrast to the side-by-side, coexisting loops in 3-space for the gyro1s. This second configuration is a gyro of the second kind, and it is designated a gyro2. A gyro2 is associated with electromagnetic waves.

Electromagnetic waves

The movement of a gyro2 is accomplished by the folding-unfolding chamber's cross sectional areas in each of the two quantized loops (doughnuts) simultaneously unfolding in the same direction respectively through the two seas of ether pocketons. As viewed in three-space this forms a right circular cylinder having two identical subparts where each one of these subparts is composed of empty space in its respective subpart of ether space. This neutral force-carrying particle is the photon. In ordinary movement, during the folding phase, the two quantized loops are re-created at the leading end of the photon in reversed positions. This makes both the associated electric and magnetic fields alternate their orientations as the gyro2 is moved ahead. Gyro2s are moved at an average speed of c, and there is no ether-wind disturbance of the ether space. Some other properties of gyros of the second kind are given below. (These properties are all established in Chapter 10.)

***E* = *m c*-squared**

Using gyro2s the energy of each photon can be determined by noting that the unfolding process forms the photon at a linear velocity of $2c$. If the quantized mass of this wave's gyro2 is m, then for this unfolding phase the apparent kinetic energy is half of m times ($2c$)-squared. During the half of time of the folding or filling phase there is no movement. Averaging these together gives the kinetic energy of the photon with rest mass m to be m c-squared. A related proof holds for gyro1s and then by heredity for the nuclei of all atoms. Note that the average speed of an electromagnetic wave going through the sea of ether pocketons is c when measured by local rods and clocks at rest in zero gravity.

Shape of a photon

Light will have the proper frequency and wave length, if the radius of the circular cross-sectional area for each of the two identical subparts of any given photon is proportional to the quantized energy of the given light ray and the length of the photon is taken to be half of the given light ray's wave length. That is, the photon's radius is proportional to the number of nuggetrons (or pocketons) that are unfolded per unit length, and the length of the photon is the linear length that this cross sectional area travels in the sea of ether pocketons during one complete unfolding. This length in turn is half of the wave length of the given light ray. In unitivity theory this configuration is used to prove that the quantized energy for any given light ray is Planck's constant times the frequency of the given light ray. (See Chapter 10.)

Quantized mass particles and force-carrying particles

Note that any given electromagnetic wave is quantized in two ways. One way is in its mass state as two, matched quantized loops, and the other way is in its wave state as a quantized shaped absence of pocketons in each of the two subparts of ether space. This explains why there are two ways that light can be reflected. One way is to change directions in the photon state on the front of the reflecting surface. The second way is to change directions in the quantized loop state or gyro2 state on the back side of the reflecting surface. In the first case half a wave length is lost from the returning wave and in the second case the wave returns in phase. Density gradients in the sea of pocketons explain why these events take place on the respective sides of the reflecting surface. (See Chapter 10.)

Doppler effect

The Doppler effect of light is determined by the way photons are filled in the folding phase. The photon is filled until the density around the folding chamber at the leading end of the photon is equal to the density of the surrounding ether pocketons at this leading end. This means that waves are repeatedly under filled when coming out of a given gravitational field and in this case lose energy and matched anti-energy. If the gravitational field happens to be moving in the opposite direction to that of the wave, this loss is increased. If the gravitational field happens to be moving in the same direction, this loss is decreased.

Other properties of light

Here are a few other properties of light.

1. The bending of light happens when the pocketon's density gradient is at an angle to a given light ray's direction of movement. The photon is shorter on the dense side.

2. Blue light may be turned to red light when two rather short blue light photons merge in length, and then split apart to form longer photons of red light.

3. When a reflection of light takes place in only one of the associated quantized loops, then a gyro of the second kind can be converted to a gyro of the first kind. The light ray goes from traveling at the speed of light to being at rest in the sea of ether pocketons where it is forced to move as a magnet by employing the nudging process in the sea of ether pocketons. A half reflection of a gyro of the first kind can return it to a gyro of the second kind. This phenomenon makes it possible to stop a light ray and then get this same light ray going again.

4. The energy of a gyro2 can be stored in a stranded photon which has had it associated folding chamber destroyed. This energy can be picked up again when the folding chamber of some force-carrying particle has become intermingled with this stranded vacuum energy. Any such stranded force-carrying particle is called heat. Note that there is positive heat in an empty hole in positive ether space, and negative heat in an empty hole in

negative ether space, and if the two holes carry the same quantize number, the two matched holes are neutral heat.

Heat

Any force-carrying particle that has lost its folding chamber is simply a stationary vacuum hole in the sea of ether pocketons. This vacuum hole is called heat. The energy associated with this vacuum can be harnessed by any active folding chamber that happens to come in contact with it. When this happens additional folding will be required to fill the combined vacuum holes. This in turn means that additional mass is folded and there is an additional pushing force. This explains how heat makes objects expand and become more energetic.

The rather isolated circuits of a swirl or quark tend to separate the gluons from heat. However, in the case of intense heat, such as exists in an atomic bomb, a chaotic change can take place in the gluons causing even atoms to break apart.

Relativity and unitivity

Relativity and unitivity agree on some things. Unitivity is a theory of everything, and as a complete theory, contains a far greater description of the structure of the universe than that found in relativity. The two theories agree that $E = m$ c-squared, and at the gyrol level, that the Lorentz transformations hold for the changes in mass, time, and length as a function of velocity. However, they disagree as to what velocity is to be use in these transformations. Relativity uses the velocity relative to another body, which is taken to be at rest in empty space. Unitivity uses the velocity of a single body going through the sea of ether pocketons.
Unitivity also, moves bodies by having them carried along in moving gravitational fields. This velocity modifies the speed at which bodies are moving relative to each other as viewed from a housing Euclidean space, but this velocity does not enter into the Lorentz transformations. Interestingly, moving gravitational fields greatly reduce the speed of a "large" body's movement through ether space when circling around an even larger body, and this, along with the rendezvous phenomena makes the constancy of the speed of light approximately true for these "large" bodies. The larger the bodies are, the better is the approximation.

The main difference between relativity and unitivity is that unitivity uses three dimensional string points to yield an ether space with two, four-dimensional subparts, while relativity has time as the fourth dimension and an empty three-space with only one part. The two subparts of ether space in unitivity are needed in order to conserve action and reaction, to conserve energy, to produce a correct unified field theory and to have two domains in which the creative-destructive activity can operate to produce two different charges. Two symmetrically equivalent, ether type sub-worlds give a structure for the universe that is far richer than that of a world with only a one-part, empty space.

Relativity theory does not conserve energy and momentum in the formation of the universe or even when making some calculations within the universe. Considering all the work that has been done working in the space of relativity without discovering a theory of everything, one must conclude that, evidently, the space of relativity is too thin to be capable of carrying a theory of everything. (Relativity must be missing, or at least not be using, at least one independent axiom in its definition of our universe.) Unitivity theory confirms that the conservation of energy and momentum and the possession of an activity are fruitful for defining our universe and are vital in obtaining a theory of everything.

5 Relation of Light, Quarks, and Time

Unitivity theory reveals the inner truths of the universe. Things too small or too strange to be observed by employing only vision are made manifest using the art of reason and independent basic laws or axioms. In order to become more familiar with unitivity theory some of the truths it reveals will be listed.

1. An electron and an associated positron come paired and these two doughnuts coexist as a single doughnut in 3-space. This neutral particle is the basic building block for quarks. It has been given the name "gyro of the first kind" or for short, a gyro1.

2. A gyro1 is a "small" magnet. There is a north pole on one side of a gyro1 and a south pole on the opposite side. When the gyro1 is not being moved through ether space, each of these poles exists around a point on the axis line which goes through the center of the coexisting doughnuts perpendicular to their common face.

3. A gyro1 in its mass state is always at rest in the ether space of unitivity theory. But a gyro1 is not always in its mass state. In fact it exists only by continually coming and going. In its mass state it is just an electron and a positron, but this existence is only for a brief instant. By a process called "unfolding" the electron and positron of the gyro1, acting as a team, react with ether space destroying themselves and a proportionate amount of the ether. The creative-destructive activity of unitivity theory conserves energy by simultaneously producing and simultaneously taking away, both energy and a canceling amount of anti-energy.

4. The most-basic building blocks for ether space are called "pocketons." They are associated with the most-basic building blocks for the electrons and positrons which are called "nuggetrons." The existence of these building blocks is required in order for the conservation of energy to be satisfied. There are positive pocketons and positive nuggetrons which are associated with the positrons. There are negative pocketons and negative nuggetrons which are associated with the electrons. The pocketons and nuggetrons fuel the energy preserving unfolding. A process called "folding" is capable of producing pocketons and nuggetrons, and consequently is capable of reproducing a gyro1. The folding process takes place in an energy conserving manner by producing both energy and anti-energy.

5. Gyrols being magnets implies that a "small" group of gyrols can be attracted together to form a line or train of gyrols. Closely lined up gyrols which are alternately, completely out of sync will have a north pole at one end and a south pole at the other end of this lineup.

6. A train of alternately out-of-sync gyrols that loops back to its beginning end forms the building block for a quark, and it is called a "swirl." Two properly matched swirls form a quark and an anti-quark. Also, three and possibly more properly matched swirls can form the same number of tied together quarks.

7. The swirls are held together by the coming and going of circuit pocketons using the gluons as the force-carrying particles. This is explained further when force-carrying particles are discussed below. The positive circuit pocketons orbit in one direction and the negative circuit pocketons orbit in the opposite direction. There are offsetting pulls associated with the unfolding of pocketons along with their associated nuggetrons, and there are offsetting pushes associated with the folding of pocketons along with their associated nuggetrons. The folding-unfolding activity produces the force that holds quarks together, and also, produces the forces associated with gravity, electricity, and magnetism.

8. A gyrol produces a north pole by producing extra negative pocketons at this pole, and produces a south pole by producing extra positive pocketons at this pole. In order for a pole to be formed, pocketons must be moved. This means that in gyrols the unfolding must be in the opposite direction to that of the folding in order to properly move pocketons and form poles. On the other hand to allow pocketons of each type to move and not produce a turbulent flow, the pocketons of each type must move like trains with out breaking rank. This means that pocketons of one type, when overloaded in a local region compared to the other type of pocketons, move in lines like trains move, but they do not move like flowing water. Iron filings over any magnetic field display how each of the two types of ether pocketons move. But at the present time little is known as to which one of the two types of pocketons is doing the moving along any particular given line. Possibly, in 4-space, both are moving along the same given 3-space line, but are moving in opposite directions.

9. An unfolded electron or an unfolded positron becomes an associated hole in its respective subpart of the ether pocketons. An unfolded, shaped, quantized hole, which is made-up of empty 4-space, is called a "force-carrying particle." In the case of a non-moving, single gyrol there is an

absence of positive pocketons on one side and an absence of negative pocketons on the opposite side. These two holes are force-carrying particles, and are called a "positive boson", and a "negative boson", respectively. In quarks the force-carrying particles, called gluons, are being formed and removed at the same time making the gluon a virtual force-carrying particle in that it never exists in a completed, quantized, shaped hole.

10. In the unique situation where a gyro1 has either just its associated positron or just its associated electron become a boson by unfolding in a direction that is opposite to that of its normal direction of unfolding, ends up with the positive boson and the negative boson side-by-side in 3-space, but not as two coexisting bosons in 3-space. This neutral boson having two separate subparts is a photon. When the electron and positron are folded at the same end of this photon, the gyro1 becomes a "gyro of the second kind" and is denoted using the symbol gyro2. Again the electron and positron have a common plane, but they are no longer side by side, but rather in a gyro2 they exist tangentially.

11. The energy of a gyro1 is in the movement of the associated anti-mass, circuit pocketons. The energy of a gyro2 is in the movement of the associated mass nuggetrons that constitute its electron and positron. Energy is conserved in the transformation from a gyro1 to a gyro2 or from a gyro2 to a gyro1. If the mass of the combined electron and positron is m, then the energy, in each of the two associated forms of gyros (i.e. the gyro1 and its associated gyro2) is *m c*-squared.

12. A gyro2 in its mass state is the mass form of a light ray while an out-of-sync set of gyro1s make-up the mass form of a quark. The two types of gyros are related. This shows that light and quarks are related. This is basically the reason why atoms have associated, specific, colors of light in their radiation, and why atoms move as waves like light.

13. The shape of a photon is a matched pair of right circular cylinders in 3-space which are treated as a single right circular cylinder with a radius that is proportional to the quantized energy of the associated light ray. This relation leads to the equation that the quantized energy of a given light ray is Planck's constant times its frequency.

14. In order for a quark to move through ether space as a wave, each of its gyro1s must open symmetrically to remove ether pocketons from in front of the quark and replace these with ether pocketons behind the quark. The

opening of a gyro1 forces the gyro1 to fold and unfold additional pocketons. In order to conserve energy, this means that the time for one complete unfolding and one complete folding of a moving open gyro1 must take longer than when it is closed and at rest in ether space. Or in other words, the frequency of vibrating for a given quark must slow down as the quark speeds up. Or in still other words, the unit of time for a given quark must increase as the quark speeds up. The amount of change in time as a function of the change in velocity is given by the Lorentz transformations. To increase the velocity of a gyro1, the gyro1 must fold at least one more positive pocketon and at least one more negative pocketon in its symmetric folding phase. This means that a gyro1's velocity is quantized, and also, as it only moves during the unfolding phase, its movement is discrete. The quantized changes in velocities imply there are only quantized changes in the Lorentz transformations. It follows that the changes in time, mass and length at the gyro1 level are all quantized just like the changes in the color of light are quantized.

For discussion:

1. Can energy be conserved in empty space?
2. In a race can a gyro1 keep up with a gyro2?
3. How is a gyro2 changed into a gyro1? How is energy conserved in this transformation?
4. How can an electron and a positron coexist in 3-space and yet as distinct particles move alike in a gravitational field and differently in a magnetic field or in an electric field?
5. What do color and time have in common?
6. Describe a force-carrying particle and how it carries the force.
7. Debate: Gravity is black magic in the time-warped empty 3-space of relativity vs. gravity is the creative-destructive activity operating on mass in the density warped ether 4-space of unitivity.

6 Unified Field Theory

Let us continue our revelations of the universe of unitivity by developing the unified field that it possesses. Unitivity does not lead along any path that previously has been traveled rather unitivity leads where there is no path and leaves a clear trail to follow.

Blinking Space and Blinking Mass

Relativity theory is based on empty space. To introduce dark-energy and yet maintain empty space, the dark-energy is sometimes described as being out of existence, then blinking into existence, followed by blinking right back out of existence. In this empty space process, as is the case in much of relativity, there is no conservation of energy. Unitivity theory with its conservation of energy reveals that about half of mass energy and about half of the anti-mass energy are in existence at any given time. Very nearly half of the anti-mass energy associated with atoms occupies the ether space at all times, and is not blinking in and out of existence. Because, unitivity theory embraces a universe with a continually-in-existence ether space, it is quite easy to derive a unified field theory.

Gravity and the Conservation of Energy

Without the conservation of energy, gravity cannot function for long periods of time. For, if gravity requires the use of energy, it eventually will destroy all the energy of the universe. This does not happen in unitivity due to the fact that in unitivity theory energy is always conserved even in the accelerations of gravity. To produce acceleration gravity uses a source of potential energy which happens to be stored in the moving circuit pocketons that hold quarks together, plus it uses a change in time.

Space with Two Symmetric Parts

Unitivity begins with empty space, and using an energy conserving activity produces the mass building blocks called nuggetrons and the ether building blocks called pocketons. Each of these building blocks is of two types. In order to satisfy the law of action and reaction, each of these two types must be placed symmetrically into two different, but basically identical subparts of space. Each of the two types of nuggetrons must be capable of getting adjacent to their respective anti-energy pocketons, and yet each type must be stored disjoint from their respective anti-energy pocketons. All of this can be done in a very elementary way in 4-space, but does not appear to be possible in 3-space. For unitivity this is no problem, but it maybe an insurmountable problem for relativity which in essence has only three space dimensions. Time as the fourth dimension even along with higher dimensions strings is not the same as having four base vectors for spanning space with all dimensions being treated alike.

The fact that electrons and positrons act alike in a gravitational field and oppositely in an electric or magnetic field, indicates a two of some sort is required in order for this to happen. One could have a two-way switch that changed either an electron, or a positron, or both. But how one could trigger the change, how one could accomplish this change and still conserve energy, and how the fields would interact with these changed mass particles, the author does not know and will leave this discovery to the reader who thinks that it is possible.

Unitivity does not require any change in the electron or positron. Each of these always acts and stays the same. The required two is obtained by having two subparts to ether space itself.

Two symmetric subparts to space can be introduced into 4-space by putting the positive nuggetrons and the negative pocketons together and placing them only one pocketon deep in the direction of the negative, fourth-dimension axis, and at the same time putting the negative nuggetrons and the positive pocketons together and placing them only one pocketon deep in the direction of the positive, fourth-dimension axis.

Quantized Loops

To form a negative quantized loop, sew together negative nuggetrons and form an electron in the positive subpart of 4-space, and have this electron blink out of existence by unfolding in the adjacent, negative subpart of 4-space. Next, have this same electron blink back into existence in the positive subpart of 4-space at the leading end of the existing unfolded hole in the negative pocketons. Let a similar scenario hold for forming a positive quantized loop called a positron.

Density Gradient Fields

In order to complete the unified field theory, all one needs is a density gradient field in the ether pocketons in each of the two separate subparts of ether space, and for each field to have the unfolding take place in the direction of the local density gradient in the local ether pocketons. That is in the direction of maximum increase in the density of the respective local ether pocketons.

In unitivity the placing of extra nuggetrons and/or extra pocketons in a region of either subpart of ether space causes squashing. This causes the anti-mass particles, which work backwards, to pull together around this region, and to form density gradient fields of the one over *r*-squared type in the respective subparts of 4-space. The manner in which the extra nuggetrons or pocketons are places in the ether space determines whether the produced field is a gravitational, an electric, or a magnetic field. Adding a pull force to the unfolding activity, and a push force to the folding activity where each of these forces is proportional to the density of the respective pocketons, completes the development of the unified field theory of unitivity.

For discussion:

1. Compare a field formed by an object composed of neutral mass with a field formed by two oppositely-charged mass objects that are located at two different points in 3-space. Give names to these two fields that are consistent with modern physics. Do electrons and positrons move correctly in these two different fields?
2. Compare a field formed by placing extra pocketons at two different 3-dimensional points with a field formed by placing extra nuggetrons at two different 3-dimensional points. Give a consistent name to the field formed by the extra pocketons. How is this field maintained given that the piled-up pocketons are nudged back into equilibrium by being forced to move out in lines like moving trains?
3. Show how open string points in the 3-space can be used to obtain the 4-space of unitivity.

7 Seeing the Color of Light

Knowing that light rays travels at about 186,000 miles a second, one can only conclude that it should be impossible to see and determine the color of any given ray of light. This is where unitivity theory is of tremendous help. Unitivity with its mass in the mass state always being at rest in ether space, and with all mass only moving as a wave, allows a given light ray to be completely deciphered by determining its at-rest, maximum-mass shape and size.

Let us begin by completely describing what light is and how light moves. Light is a gyro2. This means that light is just an electron doughnut and a positron doughnut that in their maximum mass state are tangential in a common plane. Even though this existence during any vibration is only for a very, very brief time, it still can leave an observable imprint.

The two matched, tangential doughnuts for each given light ray have various outside radii plus various radii for their circular cross-sectional areas. The cross-sectional radius for any given light ray is proportional to its quantized energy, and the radii of its two doughnuts turn out to be inversely proportional to this same quantized energy. This means that in their mass state, a red light ray is composed of two "thin" doughnuts having "long" circumferences while a blue light ray is composed of two "fat" doughnuts having "short" circumferences. (See the cover of this book.) The cone of one's eye can differentiate between these two rays by determining at what level they fit and leave an imprint on the eye's cone.

In order for a light ray to move at an average velocity c, the linear unfolding rate through the ether must have the velocity 2*c*. This is due to the fact that any light ray spends half of its time in a stationary folding phase and the other half of its time moving in an unfolding phase. Recall that the linear unfolding produces a two-part, right circular cylinder in the sea of ether pocketons with the cross-sectional area for each of the two parts being proportional to the square of the quantized energy of the given light ray, and its length being proportional to reciprocal of this energy. Using this scenario, it easy to show that for any given ray, its quantized energy *E* is equal to Planck's constant times its frequency and that its *E* is equal to *m c*-squared where *m* is the mass of the ray's associated gyro2. It is also, easy to show that associated with a light ray's movement there are alternating electric and magnetic fields which are at right angles to each other. This is what makes light an electromagnetic wave that moves in an energy conserving way. This is all established in Chapter 10 as are the following properties of light.

Light rays can be reflected either in their mass state on the back side of a reflecting surface or on the front side of the surface in their wave state. In this later case a half wave length is lost, but in the former case the wave comes back in phase. Pocketon densities give a reason for each type of reflection to take place where it does, and with a reflection there is no change in energy. Also, it should be noted that for any given light ray, a half-reflection (that is a reflection in either the positron or the electron, but not both) changes a gyro2 into a gyro1.

For discussion:

1. How does a ray of reflected light compare with the original ray? Explain the difference between the two types of reflection.
2. Describe the relation between gyro1s and gyro2s.
3. Explain how and why blue light with its higher amount of energy than red light has a shorter wavelength than red light.

8 Questions with Unitivity's Answers

Unitivity covers the structure of the whole universe as is illustrated by the following questions and with their answers.

Question: How many space dimensions does the universe possess?
Answer: The universe has three space dimensions where each point is a point plus one line segment (positive string) starting at this point and going a short distance in the direction of the fourth dimension's positive axis, and a second line segment (negative string) starting at this point and going a short distance in the direction of this same fourth dimension's, negative axis. Thus, the universe has four dimensions, but the positive and negative line segments or strings are not visible. This means that viewing the universe from within, it appears to have three dimensions. When an object occupies a given positive string and a second object occupies its associated negative string then the two objects will appear to coexist at this point.

Question: How can one view two objects as being different, and at the same time, as being coexistent?
Answer: Mix up the batter of one object and poor it into a shaped pan in 3-space using the positive-string side of all of the points contained in the pan. Mix up the batter for the other object and poor it into the same pan, but this time use the negative-string side of all of the points contained in the pan. Now bake them together and the two objects will appear to coexist at all points contained in the 3-space pan, and yet they are completely different, side-by-side, objects in 4-space.

Question: What is charge?
Answer: The universe has four basic types of particles. Negative nuggetrons make-up negatively charged particles, and positive nuggetrons make-up positively charged particles. Neutral mass is formed by using an equal number of positive and negative nuggetrons. Note though that positive pocketons determine the length of the positive-strings in ether space, and negative pocketons determine the length of the negative-strings in ether space.

Question: How many fields does the universe possess?
Answer: The universe possesses only one field which is composed of two adjacent subparts. One subpart is composed of positive stings, and the other subpart is composed of negative strings. More dense space is

characterized by shorter strings and less dense by longer strings. There are three different ways to cause shorter-type strings to exist.

1. Local neutral mass acting as a source-point mass causes shorter strings (more dense ether pocketons) to form in both, the local positive strings, and in the local negative strings, and in this case two matched strings are equal in length. The resulting identical densities and density gradients in the two subparts of space is a gravitational field.

2. Placing negatively charged particles near one point in 3-space, produces shorter strings in this local, positive subpart of space, and placing positive charged particles at a near by point in 3-space produces shorter strings in that local, negative subpart of space. The resulting densities, and density gradients, in the two subparts of space, is an electric field.

3. Placing positive pocketons in the same manner as the negatively charged particles above, and placing negative pocketons in the same manner as the positively charged particles above, results in a magnetic field. This is obviously a unified field theory, because there is only one field that has two adjacent, independent subparts which appear to coexist in 3-space.

Question: What are some of the properties of the creative-destructive activity?
Answer: The universe is assumed to have been produced out of empty space. This demands that a creative (folding) activity exists. A destructive (unfolding) activity exists when the creative activity is run in reverse. When nuggetrons are pushed ahead into existence there is an associated pushing back of the pocketons into existence, and when the nuggetrons are pulled ahead out of existence there is an associated pulling back of the pocketons out of existence. The push and pull (the coming and going) forces on nuggetrons tend to be in the same direction and are off-set by anti-forces in the opposite direction acting on the associated pocketons. The magnitude of the push and pull forces is proportional to the local density of the relevant part of the ether pocketons. The direction of these forces on moving nuggetrons tends to be in the direction of the local density gradient in the relevant subpart of the ether pocketons.

Question: What is gravity?
Answer: Gravity is the creative-destructive activity alternately pulling and pushing on every bit of mass that is contained in each and every mass object that is located within a given gravitational field.

Question: What is a gyro1?
Answer: A doughnut electron and a doughnut positron are produced at rest in the sea of pocketons, in pairs, side by side (one doughnut in each of the two subparts of 4-space), with opposite spin rotations, and they are continually being folded and unfolded by the creative-destructive activity. The paired, side-by-side electron and positron appear to coexist in 3-space, and they form a neutral mass particle called a gyro1.

Question: What does it mean for an electron and positron to be paired?
Answer: They are formed having opposite rotational spin and they are composed of an equal number of nuggetrons.

Question: How does a gyro1 vibrate?
Answer: The creative-destructive activity continually folds and unfolds a gyro1. In 3-space, the folding activity produces positive pocketons on one side of the gyro1 and negative pocketons on the opposite side during the forming of the electron and positron. During a standard unfolding, pocketons are unfolded on the side opposite to that of their folding side. This means that the gyro1 produces a magnetic field, due to the way it moves pocketons. A gyro1 is a magnet with a north pole in the local region where the negative pocketons are in excess when compared to the adjacent, 4-dimensional positive pocketons and a south pole where the positive pocketons are in excess when compared to the adjacent, 4-dimensioal negative pocketons.

Question: What is a gyro2?
Answer: If a gyro1 unfolds all pocketons from the same side instead of the negative pocketons being unfolded from one side and the positive pocketons from the opposite side, and when the next folding takes place at the end opposite to that where the unfolding began, then the result is a gyro2. In a gyro2 the electron and positron still have a common plane, but are now tangential instead of side-by-side. If a person makes a rounded figure 8 without lifting the pencil, a gyro2 is formed with the electron on top and the positron on the bottom, or vice versa. This also, produces the proper spin orientations. Using the creative-destructive activity, a gyro2 continually unfolds a right circular cylinder hole with two subparts in the ether pocketons and then refills this hole from the leading end. This moves a gyro2 in such a way that its average speed is the speed of light. If in a gyro2, either just the electron, or just the positron, is reflected and unfolded backwards, the next folding can produce a gyro1.

Question: What is light?
Answer: Light is a gyro2 in its at-rest, maximum-mass, rounded figure 8 form or state.

Question: What is illuminated light?
Answer: Illuminated light is the creative-destructive activity repeatedly operating on a gyro2, producing a discrete movement through ether space at an average speed equal to c, the speed of light. For each vibration, the length of the quantized, unfolded, right-circular cylinder with two subparts is half the wave length of the associated light ray, and is called a photon. Note that unfolded neutral mass produces identical holes in the two subparts of space that appear to nearly coexist as one, neutral, hole in 3-space. The magnetic fields associated with light indicate that a photon's two holes, one in each subpart of space, are slightly offset. This is necessary in order to have a magnetic field associated with a photon instead of an anti-gravitational field.

Question: How is light polarized?
Answer. A gyro2 in its mass state is a rounded figure 8 in a plane. This figure 8 may be found at any angle of rotation about the direction of motion. Knocking off the ones that possess certain fixed angles of rotation, yields polarized light.

Question: How is light reflected?
Answer: Perpendicularly entering light may be reflected from the front side of a reflecting mirror in the photon state by filling the photon at its opposite end compared with its normal end for forward movement. (I.e. it is filled at the end, opposite to the direction of motion.) This results in a half wave length delay, and the cloned reflection comes back a half wave length out of phase. A nearly perpendicular entering light ray may be reflected from the back side of the glass of the mirror in its mass, rounded figure 8, state. In this case the associated electron and positron unfold in the opposite direction compared to the normal direction of unfolding. This results in an in-phase, cloned wave coming back in the opposite direction.

Question: What is a force-carrying particle?
Answer: A force-carrying particle is a shaped, quantized empty-space hole formed in one, or possibly both subparts of ether space, and it is created by unfolding a quantized number of pocketons. The creative-destructive activity produces the force associated with a force-carrying particle.

Question: Do force-carrying particles carry a charge?
Answer: Negative nuggetrons are unfolded in the sea of negative pocketons, and positive nuggetrons are unfolded in the sea of positive pocketons. A hole in the negative pocketons is a negative-force-carrying particle, and a hole in the positive pocketons is a positive-force-carrying particle. Identical but slightly separated holes in the two subparts of space obtained by unfolding neutral mass, is a neutral-force-carrying particle.

Question: How does a force-carrying particle carry a force?
Answer: The creative-destructive activity produces force on mass during the folding and unfolding of a force-carrying particle

Question: What is a swirl?
Answer: A circular arrangement of alternately out of sync, matched gyro1s is a swirl. In a swirl there is a circuit of positive pocketons effectively rotating in one direction and a matched number of negative pocketons effectively rotating in the opposite direction. One of these circuits takes an inside circular rotation and the other an outside circular rotation making them slightly different in length. In reality this is a para-quark which is highly unstable. Two swirls are properly matched when the shorter circuit of one swirl is matched with the longer circuit of a similar swirl and the joined circuits have the same type of pocketons. This combination of swirls is much more stable than a single swirl. Also, using a similar approach, there can be three or more swirls involved.

Question: What is a quark?
Answer: Two properly matched swirls make-up a quark and an anti-quark. Also, quarks can be formed using three (or possibly more) properly matched swirls. Because matched circuits cause a shifting in the charged gyro1s, the quarks appear to have charge. All the swirls used to make-up the quarks are neutral, thus, combined, matched quarks are neutral

Question: How does a quark move through space?
Answer: The creative-destructive activity orients the gyro1s that make-up a given quark in such a way that they all have a diameter pointing in the direction of the ether pocketons density gradient. (This means that the face plane of the quark is perpendicular to the direction of motion.) Next the density gradient causes each gyro1 in the given quark to swivel open about the diameter that is perpendicular to both its direction diameter and its axis of rotation. In each gyro1, when the associated electron and associated positron swivel open it follows that the two folding and unfolding

chambers swivel open. The gyrol now not only folds and unfolds pocketons in the two internal circuits, but also, folds extra pocketons in the direction opposite to that of the density gradient in the ether pocketons and unfolds extra pocketons in the direction of the density gradient. This forms a force-carrying particle on the front side of the gyrols. An unfolded gyrol in its wave state is refolded at the direction-of-motion end of the force-carrying particle. Thus, each folding and unfolding produces a discrete-step movement of the associated gyrol. Note that, because adjacent gyrols are completely out of sync, adjacent gyrols in a quark take steps alternately. This means that each quark literally walks its way through ether space in "small" discrete steps.

Question: How are gravitons and gluons related?
Answer: A gyrol's two folding-unfolding chambers can form both gluons and gravitons at the same time. When a gyrol is in zero gravity and is not being moved through the ether, the folding-unfolding chambers are only involved with a gluon which is a virtual force-carrying particle in that it is being filled at the same rate that it is being formed. When a gyrol is accelerating and moving in a gravitational field, the gyrol opens and then the creative-destructive activity is forming both gluons and gravitons simultaneously. The faster the gyrol is moving; the more the gyrol is open. The more the gyrol is open; the longer is the length of the graviton, and the slower is the vibration rate at which a graviton and associated gluon are produced and filled.

Question: How is energy conserved in a moving quark?
Answer: The cross-sectional area of each folding-unfolding chamber of a gyrol in a quark is never changed. The rate at which nuggetrons and pocketons are formed or destroyed in a given folding-unfolding chamber is invariant when measured by a clock at rest in zero gravity. This is required in order to not increase or decrease energy. All of the circuit pocketons rotating in the quark must be unfolded on each vibration. A moving folding-unfolding chamber unfolds pocketons from in front of an open gyrol and simultaneously unfolds pocketons in the circuits of the quark. In order to get both of these jobs done, the chamber is forced into taking more time for each vibration. This means that a moving gyrol must vibrate slower than a gyrol not moving in ether space. This slowing down of time is consistent with the Lorentz transformations. This is proved to be true in Chapter 10. The creative-destructive activity operates at an invariant rate for both the amount of mass unfolded per unit of time and for the linear length of unfolding per unit of time (Recall that this rate is $2c$.) independently of whether the given gyrol is moving or is not moving.

Question: How is mass increased in quarks when the quark is speeded up?
Answer: With an increase in velocity more pocketons are folded and unfolded on each vibration. This means that more nuggetrons are folded and unfolded on each vibration. Thus, there is a relation between mass and velocity. This increase in mass is also, consistent with the Lorentz transformations. Again, this is proved in Chapter 10.

Question: If *n* nuggetrons existing in adjacent gyro1s weigh 2 lbs, how much do *n* adjacent, folding and unfolding pocketons weigh?
Answer: -2 lbs.

Question: If the nuggetrons in a gyro1 with mass m are being moved at an effective velocity *v* and carry *x* units of kinetic energy, how much kinetic energy do the circuit pocketons with anti-mass *m* carry in this same gyro1 if they are being moved with this same effective velocity, *v*? (With the increase in mass neglected, this is the case when the spread angle of the gyro1 is 45 degrees and *v* is approximately .7*c*. See Chapter 10, theorem 16.)
Answer: *x* units. If this gyro1 is falling, gravity has converted half of the kinetic energy in the circuit pocketons to mass kinetic energy. The increase in mass and the slowing down of time are as given in theorem 16 of Chapter 10.

Question: What is the creative activity?
No Answer: The existence and properties of the creative-destructive activity are addressed in unitivity theory. It makes the whole universe exist and operate properly, but exactly what this activity is, is left open to be addressed under the heading of theology or philosophy.

9 An Analysis of Relativity Theory

The premise for this analysis of relativity theory is that the conservation of energy is always true in every aspect of the universe. Unitivity theory is obtained using this law and has this law incorporated into it. Because of this fact, unitivity theory also will be employed in this analysis of relativity theory.

The first fact that will be established is that relativity is "close" to a limit form of unitivity. Because this is true, the two theories are related, and one might conclude that the two theories are also "close." But this is not true as is often the case when a limit form is involved.
In one sense a limit is an approximation of any "near by element" that is involved in the limit process, but at the same time the limit form may have properties that are far different from those of the associated "near by elements."

When taking a limit three things may happen. A property that is true for every object involved in a limit process (1) may still be true for the limit form (2) may not be true for the limit form, or it may happen, (3) the limit form has picked up a completely new property that was not true for any of the objects in the limit process.

The properties: 1. a number is greater than or equal to zero, 2. a number is greater than zero, 3. a number is less than or equal to zero, are respective examples of these three cases for the limit form of $1/n$ as the positive integer n becomes large without bound. Note that the inequality $1/n > 0$ holds for every positive integer n, but the limit of $1/n$ as n becomes large without bound is 0.

Relativity "close" to a limit form of unitivity

If one does not appreciate taking the limit that follows, one can arrive at the same conclusions by merely setting to zero the lengths of all line segments in the string points in unitivity's ether space. This moves one from the 4-space of unitivity to the empty 3-space of relativity.

To obtain the space of relativity as a limit of the ether space of unitivity, take the limit of each of the line segments in any given string point as all of the mass in the universe and possibly "much" more is placed into one large mass object forcing it to become large without bound. When taking this limit, it is assumed that the local density of the ether pocketons is capable of becoming large without bound whenever the local mass acting as a source-point mass becomes large without bound. Note that the limit taken here is a theoretical limit only. In reality, when the amount of mass of a given body becomes large without bound, one encounters such things as imploding stars, etc.

As viewed from a housing Euclidean space, the length of each of the line segments forming any given string point is equal to the length of one side of an adjacent pocketon, and this length can be made shorter than any arbitrarily small positive number epsilon by merely making the density of the pocketon sufficiently large. The density of pocketons is directly proportional to the amount of mass that is acting as the source-point mass. Thus, as the given body's mass becomes large without bound the limit of the length of each line segment of an arbitrary string point is zero.

Taking this limit has moved the space of the universe from being that of unitivity's four dimensional ether space having two subparts to being that of relativity's three-dimensional empty space with only one part.

This obviously has a number of implications. There is now no movement through ether space. Unitivity's light waves and de Broglie waves are now gone. This means that the constancy of the velocity of light is now in the realm of the possible, but this also, means there is no unique and universally consistent velocity left to be used in the Lorentz transformations, nor to be used for the conservation of energy and momentum. All movements are now strictly relative.

Note that this limit was attained by employing a body that is large without bound. Obviously, then this limit form can be approximated by using "large" bodies in empty space. The taking of this limit gives one path from unitivity theory to relativity theory.

Let us review the properties that are preserved and the properties that are lost. One property that is true in the ether space of unitivity and is preserved in the limit space of relativity is $E = m$ times c-squared. One property that is lost is that of having a space with two parallel ether subspaces, and because of this loss, the unified field theory and TOE of unitivity are now gone. Another property that is lost is the unique velocity for any object moving through ether space. Because of this loss, a unique velocity to be used in the conservation of energy, momentum and the Lorentz transformations is now nonexistent. One property that is picked up is that all movements are now strictly relative and as stated above, with this comes the possibility for the constancy of the velocity of light.

Comparison of unitivity and relativity

The preservation of $E = m$ c-squared for light rays that was preserved when the limit was taken is good, but it does not necessarily imply that all of relativity theory is true.

Unitivity theory reveals that the energy in an atomic mass m not moving through ether space is stored as an equal amount of anti-mass m

that effectively is traveling at the velocity $2c$, for half of the time. This means that this anti-mass has a kinetic energy equal to m c-squared. (Actually, this is the reason why $E = m$ c-squared is true for neutral mass.) The creative-destructive activity operating on a gyro1 in a quark that is in a gravitational field is capable of producing a change in the gyro1's kinetic energy by reducing the kinetic energy of this associated revolving anti-mass. Also, in this exchange of kinetic energies there is no change in the total energy as is proved in Chapter 10. In unitivity theory, kinetic energy is carried in discrete steps by the force-carrying particles. Mass in its mass state is always at rest in the local ether space. A given amount of mass is moved as a wave by being destroyed and becoming a number of force-carrying particles which in turn are turned back into mass in the mass state. The force-carrying particles can be shown to always conserve energy and momentum because of the manner in which the creative-destructive activity operates within them.

It should be noted that in order for the creative-destructive activity to accomplish this exchange of kinetic energy and at the same time conserve energy, both mass and anti-mass must be increased, and time must be slowed down. How this all works is established in Chapter 10. Unitivity theory demonstrates that it is possible to conserve energy during gravitational accelerations, but note that this can be accomplished only when the required exchange energy is already in existence. This in-existence source of energy is never discussed in relativity.

In the above limit form there is no velocity of objects moving through ether space. This means that this velocity must be replaced by a velocity which is relative to something else. The obvious choice is to use a velocity relative to another object as is done in relativity. According to unitivity, this means that the velocities of moving gravitational fields and the velocities of objects moving through ether space (both of these as viewed from a housing Euclidean space) are now replaced with velocities of only one type, namely relative velocities. This leads to a number of interesting difficulties.

The conservation of energy vs. relative motion

In unitivity theory, energy and anti-energy along with the law of action and anti-action working together produce a sum total energy for the whole universe that continually has the value zero. It will now be shown that this is not the case when all movements are simply relative movements in empty space. For the conservation of energy considerations in relative motion, let us simplify the argument by assuming the whole universe consists of only two individual bodies which are moving relative to each other where each is large enough to be a rough approximation to being large-with-out-bound. Further, assume that this particular universe possesses a total amount of energy, X.

With only relative motion it is not possible to conserve this amount of energy X nor is it possible to conserve momentum for these two arbitrary, moving bodies. To see this, let these two bodies have mass m and M respectively with m not equal to M. Let the bodies be moving apart at a velocity V which has the magnitude, v. Depending on which body is considered fixed, the kinetic energy is either half of m times v-squared or it is half of M times v-squared. When no other changes in energy are made, and when m and M are not equal, the energy is changed whenever the fixed point is moved from one body to the other body. Thus, it is not possible for X to have a single value. A similar argument holds for momentum.

Even more significant is the fact that when two points are moving with respect to each other and both are taken to be fixed in the same discussion with no adjustments made to permit this, then the conservation of energy has been violated and two different energies are now incorporated right into the resulting theory.

This fact alone means that relativity theory is at odds with our "real" universe to anyone who is convinced that the laws of the conservation of energy and momentum must always be satisfied. This is where an ether space becomes an absolute necessity. It is necessary for determining a true and everywhere consistent base velocity, which in turn can be used to conserve energy and momentum and can be used in the Lorentz transformations. Note too, that ether space is needed for traction, i.e. to have something to push and pull on when conserving momentum in gravitational accelerations. Also, ether space is needed to hold the dark-matter energy which according to unitivity theory consists of the mass state of force-carrying particles. (See Chapter 11.)

In the discussion that follows, the existence of an ether space is considered to be an absolute necessity in order for energy to be conserved. Let us review the above remarks in order to put a greater emphasis on them. For the force of gravity to conserve energy, there must be a local velocity through the ether which is well defined and which can be used to determine the local kinetic energy of any atom moving through ether space. Gravity must have an associated activity that is capable of transferring kinetic energy already in existence to kinetic energy in falling atoms. Further, the activity must make this transfer without using up any energy on itself, and it must use ether space for traction in order to not violate the law of action and reaction. As noted above this is what happens in unitivity theory. However, relativity with its empty space has no way of offsetting any gravity produced acceleration and no way for uniquely measuring any change in the velocity through space. Consequently, relativity is forced to neglect the conservation of momentum and energy as they relate to the whole universe.

Newton was completely convinced that an ether space existed. He made great advances in the knowledge of gravity and light based on this conviction. He also, was convinced that light had an associated particle. He was aware of matter, and according to unitivity, his thinking that light had a particle or a mass state made him aware of cold-dark matter. In order to know the rest of the story about gravity and light, he needed to have access to modern day knowledge about dark-energy, about quantized light, about today's sophisticated mathematics, etc. Or according to unitivity theory, he needed to know the construction of the force-carrying particles, the role played by the anti-mass particles that are called pocketons, the existence of the creative-destructive activity, etc. Note that dark-energy in the forms of pocketons and force-carrying particles are an integral part of unitivity theory. (See Chapter 11.)

"Real" changes vs. "apparent" changes

To understand the relationship between "apparent" changes in time and "real" changes in time as given by the Lorentz transformations, let us consider the following situation. Let two objects be moving in opposite directions away from a given fixed point in ether space with velocities V and $(-V)$ respectively. Then, as viewed from this given fixed point, there will be an "apparent" slowing down of time onboard each of these two objects as they move away from the fixed point even when there is no onboard real slowing down of time. In this case the relationship between the time at the fixed point and the "apparent" times of the moving objects is given by the Lorentz transformation which employs respectively the square of the magnitude of V or $(-V)$. Again note that here we are assuming that the "real" time remains unchanged onboard the two moving objects, and that their "real" times are the same as the "real" time at the given fixed point.

Let the two moving objects that started together be simultaneously stopped and have their velocities reversed. On their return trip to the starting point there will be an "apparent" speeding up of time as observed from the given fixed point. If the factor that yields the "apparent" slowed down time is f1, then, keeping the same velocity, it is easy to show the multiplicative number, f2, that yields the "apparent" speeded up time on the return is given by $f2 = 2 - f1$. A Lorentz transformation gives the multiplicative factor f1 for obtaining the "apparent" slowing down of time. Thus, this can be used in turn to obtain the transformation for the "apparent" speeding up of time. To make a long story short, for time as given here, when these two given objects return to the given starting point, all clocks will still be in sync. The "apparent" changes are just that, they are merely "apparent." There are no "real" changes in time in this given situation. Also, here a Lorentz transformation is used only in the case of an "apparent" slowing down of time. A different, but related transformation must be used in the case of the "apparent" speeding up of time for the situation studied here.

Note that here there is no equality between "apparent" time and "real" time. They are different. In fact, when there is also, a "real" slowing down of time taking place onboard an object departing from a fixed point, then the "apparent" slowing down of time is modified by this onboard "real" slowing down of time. This means that the "apparent" slowing down of time is always greater than the "real" slowing down of time, and that whenever both are taking place in the same discussion the two can never be equal.

In unitivity theory, for the case of "large" and "isolated" moving objects as viewed in a housing Euclidean space, the moving gravitational fields account for most of the movement. Associated with this velocity there are no "apparent" and no "real" changes in time for objects moving away from each other. This movement is caused merely by expansions and/or contractions of the associated ether pocketons that make up ether space. This has no effect on the time required for a light signal to travel between two objects.

In unitivity the movement of any entity through ether space is always in terms of only the number of pocketons the entity has linearly moved past. This type of velocity, where objects are moving through the ether space, possesses both an "apparent" slowing down of time, and a "real" slowing down of time. In unitivity theory, movement causes light signals to travel linearly through a continually increasing number of pocketons as they travel between a point at rest in the local ether space and an object departing through the ether space. Thus, with movement there will be an increasing delay in the return signal making time "appear" to have slowed down. In addition, in unitivity theory this same velocity causes a "real" change of time in that the gyros of the first kind vibrate slower as their speed of moving through the ether space increases. Interestingly, because of the structure of the gyros, this "real" change in time does happen to be equal to the associated "apparent" change of time as computed above in the first example. Recall this is provided that for computing the "apparent" change in time, the onboard time is assumed to not be changing. Each of these time changes are given by a Lorentz transformation using for the argument the magnitude of the velocity of departure from a given fixed point in the local ether space.

The gyros of the first kind produce the Lorentz transformations which in turn give a weak form for the constancy of the velocity of light. When the relative speed (c-v) is mapped to an onboard c, then it is true that the total "apparent" change in time for an object departing away from a fixed point is greater than the "real" change in time. However, if one assumes the weak form of the constancy of the velocity of light is true, and uses the Lorentz transformations to compute the usual "apparent" change in time for a body departing from a given fixed point, it follows that the "apparent" change in time agrees with the "real" change in time.

If one assumes that the strong form of the constancy of the velocity of light holds, then there are two problems that need to be addressed. One is what is to be done with the "apparent" speeding up of time when two bodies approach one another, and two, what transformation replaces the Lorentz transformations in this speeding up of time case? Let us look into these problems.

One base fixed point vs. two independent, basic fixed points

To begin with the author relates the following story about his grandfather. His grandfather used a steam engine to break virgin sod on the western prairie during the homesteading days. The steam engine pulled plows with a total of 16 or more bottoms, but it pulled the plows very slowly. His grandfather always said jokingly, "The only way we could be sure the steamer was moving was to pound a stake in the ground and keep measuring the distance between the steamer and the stake."

Had his grandfather used two stakes, one pounded in the ground and one fixed to the plow, he would have been very confused. According to one of these stakes the steamer would be moving, and at the same time according to the other stake it would not be moving. He would know that it must be either moving or not moving, but he would not be able to determine which one of these two possibilities was really true. He would not be able to decide whether to start fixing the machine or to just leave it alone.

To further illustrate the difference between one and two fixed points, let us again consider two objects which are moving in opposite directions with each going away from a single given point which is at rest in the local ether space. Let each have the speed v going through the ether space. According to unitivity, on board each of these objects the time has slowed the same amount and is obtained using v for the argument in the Lorentz transformations. But note that the two objects are moving apart at a speed 2 times v. If the speed $2v$ is used in the Lorentz transformations to obtain the "real" on board slowed down time, then "the calculated value for real time" is slower than the "real time as recorded by the moving clocks." This again illustrates that "apparent" and "real" changes in time are not one and the same thing.

When only one fixed point of reference is used, the use of the speed v for the argument in the Lorentz transformations is indicated. When two fixed, and at the same time moving, points are used in the derivation, as is done in relativity, then the use of the speed of departure, $2v$, is indicated. Which one of these values is correct?

It is not logically possible to have a point at rest and moving in the same discussion. Logic does not allow a statement and the negation of the same statement both to be true in a single calculation. The relativity procedure that introduces into equations the velocity of objects by using two separate and independent fixed points, without making any modifications in any existing equations, can result in a value for the argument of the Lorentz transformations that is twice as large as that obtained when using only a single fixed point.

Let us return to the question, "What happens to the 'apparent' speeding up of time for two objects moving toward one another?" When two objects move apart and then are brought back together, the two different "apparent" changes in time tend to, and in the case where the magnitude of the velocity is kept constant, do cancel one another. However, according to unitivity theory, the velocity through ether space does cause a "real" slowing down of time both going and coming back. Relativity states that this is true also, but fails to address the "apparent" speeding up of time when two objects move toward one another. This "apparent" change in time is just as "real" as the "apparent" slowing down of time when the objects are moving apart from one another. The fact that "apparent" slowing and "apparent" speeding up of time are equally "real" means one can never just switch from "apparent" to "real" changes in time without carefully explaining exactly how this is being done. If body #1 is moving away from body #2 and toward body #3, then switching "apparent" for "real" would mean that the "real" time on body #1 would have to both slow down and speed up at the same time. This is impossible. This type of problem does not arise in unitivity with its ether space.

In unitivity theory, light waves and de Broglie waves have the relative velocities $(c+v)$ and $(c-v)$ as viewed from a housing Euclidean space. The gyros of the first kind produce the Lorentz transformations that map $(c-v)$ to c whenever the gyros are moving at a speed v through the ether space, and the value $(c+v)$ is mapped to a modified $(c+v)$ value.

The reason this modified $(c+v)$ value is hard to detect is due to the moving gravitational fields and the rendezvous phenomenon. As viewed from a housing Euclidean space, the velocity of "large" mass objects going through ether space tends to be an extremely small percentage of the total velocity. Consequently, the difference between the velocities of c and of $(c+v)$ tends to be very small and hard to detect by anyone who is checking this on a moving "large" mass object. In addition the rendezvous phenomenon makes the detection of movement through ether space even harder to detect. (See theorem 24 in Chapter 10.)

The conservation of energy added to relativity?

What changes are required in order to introduce the conservation of energy into relativity theory?

1. The use of two, independent, fixed points must be replaced with a single, independent, fixed point in any given discussion.

2. An ether space must be introduced in order to have a fixed point and unique base for measuring velocity.

3. Gravity must be given an associated activity that has the capacity to impart precise amounts of kinetic energy to any given atom by reducing some other, in-existence energy by this same amount, and further, this must be accomplished without the activity using up any energy on itself or violating the conservation of momentum.

4. "Apparent" changes in time both of the slowing down and the speeding up types must be addressed as different types, but as equally important. The use of the Lorentz transformations must be used to map only a "real" relative velocity of $(c-v)$ to c and not take for granted that these transformations also, map $(c+v)$ to c which, according to unitivity theory, they do not.

Conclusion

In the above statements the conservation of energy and the theory of unitivity are considered to be true and have been used to reveal that relativity is related to unitivity, but relativity has some difficulties and short comings. Unitivity's axiomatic derivation with its incorporated laws of the universe avoids the problems listed above for relativity and has many additional desirable properties which have not been, and probably cannot be, established in the present form of relativity theory. A few of the missing desirable properties are: the possession of the conservation of energy and momentum, the possession of a unified field theory, the possession of two states for any single ray of light, (one that is moving fast and one that is stationary and can be seen), the possession of force-carrying particles that are quantized, the possession of an ether space that is strong enough to make galaxies obey gravity, the possession of an ether space that has the proper proportion of dark-energy (See Chapter 11.), the possession of a structure for mass that verifies that E = mass times c-squared and that this energy can be used by gravity to accelerate mass, the possession of a TOE, and the possession of an activity which consumes no energy, yet produces the forces associated with gravity, electricity, magnetism, and momentum.

The above analysis gives a great deal of credit to the genius, Einstein, because it reveals the unusual insight he possessed. Though, according to unitivity, he was working in a very elementary space that was missing essential basic independent properties, and though he was using questionable logic when he used two different points that are moving with respect to each other as being two fixed points in the same discussion, and though he was stretching the conservation of energy and momentum, he still was able to obtain a number of very new and important discoveries about our universe.

In conclusion, relativity and unitivity do not possess the same properties and as such are only partially compatible theories. The reader should use experimentation, logic, mathematics and the laws of physics to decide which of these, or any other theory, best describe our "real" universe.

Part Two
UNITIVITY FROM A MATHEMATICAL PERSPECTIVE

Including Relevant Experiments

10 A Formal Axiomatic System for Our Universe: Historical Development

The reader may like to see some major changes in the way the material is presented in this chapter, but to do this the author would have to make major changes in the way this material was originally presented. To the best of the author's knowledge, the formal axiomatic system presented here is the first formal axiomatic system to have phenomena consistent with the phenomena of our universe. In order to preserve this unique, and in a certain sense historic document, the printed material presented here has been obtained by making only limited and small changes in the original document.

Introduction

The goal in this chapter is to produce a consistent set of axioms which has our real universe as its model. The formal axioms that are introduced deal with the existence and fundamental properties of undefined quantities. The exact construction of these entities and the exact mechanism of their interactions are left open by being undefined. However, the introduced properties given to these abstract quantities produce an abstract universe that has physical phenomena that match the phenomena of our real universe. The axioms that describe the undefined terms are logical in that objects only react with adjacent objects. There is no force at a distance. Active mass interacts with a non-empty, dynamic ether space to produce the effects of gravity, magnetism, electricity, and momentum. Energy and momentum are always conserved. The changes in mass and time required by the Lorentz transformations turn out to be produced by certain mass particles acting as analog computers. In the presentation given here there is no solid, inactive mass that is influenced by a completely empty space.

At the completion of this formal axiomatic system that defines an abstract universe which has the same phenomena as our real universe, a real model for this axiomatic system is developed using a material axiomatic system. (In this book this model is developed in Chapter 2.) This real model provides an abundance of new knowledge about the inner workings of our universe.

Few axioms explain many phenomena

Surprisingly, a very small number of axioms explain a much larger number of physical phenomena. This fact lends credence to the statement, "These axioms do represent reality." The axioms give insight into many of the hard questions of physics. Also, they suggest a number of experiments which can be carried out to test the validity of the results that are obtained. However, by the same token, the axioms stated here may not produce the whole story. There may have to be some additions and/or slight modifications in order to cover rare physical phenomena of the universe.

Does ether exist?

For many years, scientists have tried to answer the question, “Is there, or is there not, an ether space that carries light, magnetism, electricity, and gravity?” Michelson and Morley’s experiment seems to have proved that there is no observed ether wind. (Others are still attempting to prove that ether drift does exist by using various techniques.) Some have interpreted the Michelson and Morley’s experiment to mean that there is no ether. However, this is not necessarily a correct interpretation. One can only conclude that mass moving through our universe’s ether space moves in such a way that the ether is left essentially undisturbed, and for some reason movement is undetectable by the Michelson and Morley’s experiment. Moving gravitational fields and the rendezvous phenomena introduced below in theorem 24, show that this is possible even within an ether space.

Mass moving through ether space

It is hard to conceive of any way the ether can be left undisturbed (i.e. left without a path-clearing flow of ether around moving bodies) other than by the following procedure:

Let a “small” set of mass nuggets and an adjacent set of an equal number of ether space pockets completely destroy each other. Following this, let them form again just as before, but with their respective positions interchanged in the ether space. A sequence of directed interchanges of this type can move a “small” set of nuggets a great distance, at a great velocity without disturbing the ether.

Two intermingled parts to ether space

A second fact to be considered when discussing any ether space is the fact that positively charged particles and negatively charged particles act alike in a gravitational field, but oppositely in a magnetic or electric field. There are three basic ways this can happen when one is restricted to no more than two things existing between the particles.

1. There is nothing in the space between particles, but there is force at a distance with some entity moving the two types of charged particles alike sometimes, but switching to move them oppositely at other times.

2. There is one entity between particles with differently charged particles reacting oppositely to this entity in an electric or magnetic field, and yet reacting alike to this entity in a gravitational field.

3. There are two separate, but adjoining and intermingled entities operating between objects in space. Particles having one charge interact with one of the two parts, and particles having an opposite charge interact in a similar way with the second part.

It is hard to imagine exactly how (1) or (2) can possibly satisfy all the necessary requirements for proper movement of charged particles and at the same time support a unified field theory. When two different objects react differently to two different fields, it is difficult for the two different fields to be essentially one field unless the field has at least two separate parts. Further, a symmetric treatment of both positive and negative particles along with a symmetric treatment of mass and space favors (3). In the following axiomatic system, (3) is taken to be true. It then turns out that saying some entity carries the force as stated in (1) above, is analogous to saying an air bubble rising in water is the carrier of the force, when, in reality it is the pressure gradient in the surrounding water that is producing the movement.

Two Fundamental Concepts

Movements where objects are destroyed and then reformed at a new position, plus a space, which possesses two independent parts, are two fundamental concepts used in forming the following axiomatic system which has our universe as a model.

Influence-Space is Designated I-Space

To be consistent with curved or warped space terminology, the words ether space for the most part will be used rather than just the word ether. The special space and special mass that may influence each other to disappear and then to reappear, shall be designated I-space and I-mass, respectively. The "I" here refers to the word influence. This means that I-space is an influence-space and I-mass is an influence-mass. They influence each other to unfold in folding-unfolding chambers. All of the properties of undefined I-space and its associated undefined I-mass are stated as axioms. Some of these axioms are justified by observations. Others are justified by the fact that they are needed in order to produce phenomena consistent with scientific observations made in our universe. Within any formal axiomatic system, all the stated axioms are considered to be true. They do not need to be proved, and within the axiomatic system, one cannot argue with the given axioms unless they happen to produce an inconsistent system.

In the following axiomatic system, the letter "I" is placed also in front of many physical entities. This is done to emphasize that the axiomatic system produced here is itself abstract and that it refers to an abstract universe. Also, having the "I" in front of terms that have meaning in our real universe makes it easy to eventually define these abstract terms using real terms. Undefined terms must be defined in order to produce a real model for an abstract, formal axiomatic system. (In Chapter 2 of this book a real model for the following formal axiomatic system is produced using observed material axioms, and the real model produced should be our universe.)

Overview of the Formal Axiomatic System's Implications

A quick overview of some of the highlights and implications associated with this axiomatic system are listed below. (In this list all leading I's have been omitted.)

Illuminated light forms holes (or flaws) in ether space. These holes are quantized and are basically right circular cylinders, where each cylinder has two identical sub-cylinders that have a constant cross-sectional area that is proportional to the square of the energy of the associated light ray. From these facts it follows that

$$E_o = h\nu .$$

(Here E_o is the quantized energy, h is Planck's constant and ν is the frequency of the given light ray.)
Another descriptive way to state this is as follows: Let ℓ be distance, t be time measured by a rod and clock at rest in uniformly dense space, and let c be the speed of light in this uniformly dense space, then when

$$\frac{d\ell}{dt} = 2c$$

and

$$\frac{dE}{d\ell} = -\frac{2}{ch} E_o^2 \text{ with } E(0) = E_o \text{ and } E(\tfrac{\lambda}{2}) = 0$$

(here $\lambda/2$ is half the wave length), it follows that

$$\nu\lambda = c$$

and

$$E_o = h\nu .$$

The extra 2's seem strange, but for a complete cycle the energy E_o must be unfolded and folded twice, and there is only effective movement during the unfolding phase and this phase only accounts for half of the involved time. The electromagnetic property of light, the bending of light, the polarizing of light, the extreme heat of a laser beam, etc. are each proved in a theorem by using the axioms of the given axiomatic system.

Similarities and differences of magnetic fields and electric fields are illuminated.

The relationship between gravity and force-produced accelerations is illuminated, including the reason a karate punch is so effective.
An atom's unique binding energy is illuminated along with the unique, associated energy of its orbital electrons.

At the gyro-1 level, in the direction of motion, a length L at rest will be changed to $L \cdot \sqrt{1-\frac{v^2}{c^2}}$ when the velocity through ether space is v. The same opening of the gyro-1s that causes this change in length causes objects made from gyro-1s to maintain the velocity v until acted on by a force.

The fact that all movement of any given gyro-1 depends on the vibration rate for a complete unfolding of the given gyro-1 and associated space, followed by a complete folding of the given gyro-1 and associated space, leads one to the conclusion that mechanical and crystal watches speed up or slow down together in sync as their common velocity decreases or increases respectively.

The way atomic mass increases with an increase in velocity is illuminated, along with the reason why the velocity of light is an upper bound for the velocity of any mass object moving through ether space. Additionally, expanding and contracting ether space is shown to carry mass objects, too.

An increase in velocity through I-space causes a shortening of the length of gyro-1s in the direction of motion, an increasing of the mass of these gyro-1s, and a slowing down of time, that all agree with the theory of relativity. However, there is one big difference between relativity and the unitivity theory produced here. In I-space you cannot interchange the roles of a system at rest and a system in motion. Consequently, the paradox of two observers, each growing older slower than the other, does not even come-up in the unitivity theory that arises out of the following formal axiomatic system.

It is quite amazing how so many things can be explained and proved using the ten axioms that are stated below.

The Formal Axiomatic System

Anyone reading this book and not familiar with axiomatic systems is encouraged to read the definition of axiomatic systems given in Chapter 2 before going on to read the rest of this chapter.

Axiom 1. Existence of I-mass and I-space

There exists I-mass (undefined) and I-space (undefined). It is assumed that the abstract universe associated with these two undefined terms does possess defined terms like energy, momentum, length, density, velocity, Planck's constant, spin, and a consistent set of units, say gm, cm, sec, all of which can be used for calculation purposes. This is to say that the physics in the following abstract universe is assumed to be consistent with the physics in our real universe except for the undefined terms, and that the only things to be taken as undefined are those things explicitly stated as being undefined. Further, it is assumed that whenever desired, all the above can be housed in a super-imposed, mathematical, Euclidean space.

Axiom 2. Composition of I-mass

I-mass is composed of two intermingled, but distinct, finite sets of nuggets (undefined). One of these sets has all of its nuggets composed of I-neg-mass (undefined). The other set has all of its nuggets composed of I-pos-mass (undefined). The nuggets of the two different types come paired, and in any "large" region of I-space, there is approximately a 1-1 relation between them. Further, an I-neg-nugget, (i.e. a nugget composed of I-neg-mass), or an I-pos-nugget, (i.e. a nugget composed of I-pos-mass), consists of an amount of energy equivalent to half of Planck's constant (i.e. they all have approximately the energy 3.31×10^{-27} gm cm^2/sec^2).

Axiom 3. Composition of I-space

I-space is composed of two intermingled, but distinct, finite sets of pockets (undefined). One of these sets has all of its pockets filled with I-neg-space (undefined) and the other set has all of its pockets filled with I-pos-space (undefined). These pockets satisfy the following:

1) The pockets of the two different types come paired, and in any "large" region of I-space, there is approximately a 1-1 relation between them.

2) Each pocket, at any time t, has a size (volume), and any set of pockets of either type has an average size $\bar{s}(t)$ all as measured in a housing Euclidean space. (The pocket's density is the number of pockets per unit volume as measured in a housing Euclidean space.)

3) The pockets of I-neg-space and I-pos-space adhere together and tend to lock in an orderable, 1-1, intermingled pattern (undefined). (Note that they do not completely lock, and that they can move without expending any energy in a special manner called a nudging process that is to be described later. Plus they can expand or contract in an ordered array without expending any energy.)

4) In a region where there is no I-mass, and the two types of I-space pockets are in a 1-1 relation, the densities of pockets of I-neg-space and I-pos-space are equal.

5) I-mass of either type forced against the pockets of I-space will increase the density gradient of the pockets in the direction of the force, but the pockets will not permit any I-mass to pass by or through their locked together pattern.

6) Individual pockets on the edge of I-space (face of the I-universe) and next to empty Euclidean Space, expand to a maximum size (volume), $\hat{s}$, where $\hat{s}$ is a function of the local density of I-space. The pockets adjacent to the outside pockets, but just inside expand to a size $s_1(t)$ where for some $r_1(t)$

$$s_1(t) = (1 - r_1(t))\hat{s}.$$

The next layer of pockets have size $s_2(t)$ where for some $r_2(t)$

$$s_2(t) = (1 - r_2(t))s_1(t).$$

Repeating gives two finite sequences $\{s_i\}$ and $\{r_i\}$ at any region on the edge of I-space. These sequences describe a local density gradient field on the edge of I-space as a function of time. The r_i depend on the overall average size of the pockets, $\bar{s}(t)$. (For example, when $\bar{s}$ is small compared with $\hat{s}$, the density gradient on the edge of I-space will be large, the r_i will be relatively large, and I-space will be expanding into empty space. For a universe that is formed using a single mass-point source, it is shown in Chapter 2 that this universe eventually attains and maintains a density drop-off that is proportional to one over r-squared where r is the distance from the center of this universe. This infers that this particular universe is expanding when the drop-off in density is greater than one over *r*-squared and is contracting when the drop-off is less than one over *r*-squared.)

7) Pockets are compressed in the presence of I-mass and/or additional I-space, and they obey the following rules:

> a) Additional I-pos-mass and/or I-neg-space placed in a local region of I-space compresses the pockets of I-neg-space which are already there. This forms a density gradient field in I-neg-space. The density gradients' magnitudes are proportional to the amount of I-pos-mass and/or I-neg-space placed there and these magnitudes are inversely proportional to the square of the distance from the source-point entity that was placed there. The amounts of I-pos-mass and the distances are as measured in I-space. Different density gradient vectors produced at a given point from different sources are vector additive. The densities of pockets of I-pos-space are not affected in this situation.
>
> b) Additional I-neg-mass and/or I-pos-space placed in a local region of I-space compresses the pockets of I-pos-space which are already there. This forms a density gradient field in I-pos-space. The density gradients' magnitudes are proportional to the amount of I-neg-mass and/or I-pos-space placed there and these magnitudes are inversely proportional to the square of the distance from the source-point entity that was placed there. The amount of I-neg-mass and the distances are as measured in I-space. Differently produced density gradient vectors at any point are vector additive. The densities of pockets of I-neg-space are not affected in this situation.

8) In a local region of I-space where there exists a surplus of either I-neg-space or I-pos-space in comparison to the other, there develops a process of nudging (undefined, but in this nudging process, the pockets of the type that are in surplus are forced to move in an ordered array, in a direction opposite to that of the local density gradient field. That is, the movement is like trains move and the movement is away from high density areas toward low density areas).

9) The adjustments in densities of I-space are made with the leading influence of an I-gravitational-wave spreading out in all directions "nearly" instantaneously, followed by more precise adjustments taking place along reflected I-gravitational-waves. However, the nudging of pockets of either I-pos-space or I-neg-space being nudged into equilibrium tends to be a "slightly" lingering event.

10) The size or density of nuggets composed of I-neg-mass are directly proportional to the size or density, respectively, of pockets of I-pos-space adjacent to them. Also, the size or density of nuggets composed of I-pos-mass are directly proportional to the size or density, respectively, of pockets of I-neg-space adjacent to them. All of these density measurements are measured in a housing Euclidean space.

Axiom 4. Relationship between I-space and I-mass

I-neg-space is anti-I-neg-mass, and I-pos-space is anti-I-pos-mass.

This is to say that in a folding-unfolding chamber a fixed number of adjoining pockets of I-neg-space can influence the same fixed number of adjacent nuggets of I-neg-mass to unfold (undefined, but does mean, when viewed from a housing Euclidean space they both disappear leaving a hole or flaw in I-neg-space with axiom 3, part 6 applying on the boundary of this hole.) When properly triggered (undefined here, but axiom 5 does mention a triggering mechanism), and void of any outside influence, in the given folding-unfolding chamber the same fixed number of pockets of I-neg-space along with the same fixed number of nuggets of I-neg-mass will be refolded (undefined, but does mean that the unfolded pockets and nuggets both reappear). This process of unfolding and refolding leaves everything exactly like it was when the process started, except for one thing, which is, that under most conditions, the relative positions of the pockets and nuggets involved have been interchanged. The number of nuggets and the associated, equal number of pockets unfolding linearly or refolding linearly, per unit of time, as calculated using a local clock at rest in uniformly dense I-space, is independent of the density of the pockets and nuggets. A statement similar to this holds for I-pos-space and I-pos-mass. Also the above two statements imply a similar statement holds for I-neutral-space and I-neutral-mass which are defined in the following definitions 1 and 2.

Definition 1.

A body or particle of I-mass which contains an equal number of nuggets of I-neg-mass and nuggets of I-pos-mass is called an I-neutral-mass body or particle or just I-neutral-mass.

Definition 2.

A region in I-space which contains an equal number of pockets of I-neg-space and pockets of I-pos-space, in any "not too small" sub region, is called an I-neutral-space region or just I-neutral-space.

Definition 3.

A body or particle of I-mass which has a surplus of I-neg-mass compared to the amount of I-pos-mass is called a negatively charged body or particle. The measure of negative charge is proportional to the surplus amount of the I-neg-mass. A body or particle which has a surplus of I-pos-mass compared to the quantity of I-neg-mass is called a positively charged body or particle. The measure of positive charge is proportional to the surplus amount of the I-pos-mass.

Definition 4.

The density gradient field formed in I-neg-space by the presence of a surplus of I-pos-mass, compared to I-neg-mass in a local region of I-space, is an I-neg-electric field.

Definition 5.

The density gradient field formed in I-pos-space by a surplus of I-neg-mass, compared to I-pos-mass in a local region of I-space, is an I-pos-electric field.

Definition 6.

An I-pos-electric field formed by a surplus of I-neg-mass in one local region of I-space, and an I-neg-electric field formed by a surplus of I-pos-mass in a neighboring local region of I-space, is an I-electric-field.

Definition 7.

The two density gradient fields formed in the two parts of I-neutral-space by the presence of a body of I-neutral-mass in a local sub region of I-neutral-space is an I-gravitational-field, i.e. to say when "locally" the density gradients of the two I-electrical fields are in the same direction and have the same magnitude, the I-field formed is called an I-gravitational field.

Definition 8.

The density gradient field formed in I-neg-space by the presence in some "small" local region of I-space of a surplus amount of I-neg-space compared to the amount of I-pos-space is an I-neg-magnetic-field.

Definition 9.

The density gradient field formed in I-pos-space by the presence in some "small" local region of I-space of a surplus amount of I-pos-space compared to the amount of I-neg-space is an I-pos-magnetic-field.

Definition 10.

The two fields formed by a surplus of I-neg-space compared to I-pos-space in one "small" local region of I-space and a corresponding surplus of I-pos-space compared to I-neg-space in a neighboring "small" local region of I-space, along with the resulting nudging (flux) of pockets (see axiom 3, part 8), is an I-magnetic-field.

> **Definition 10.1.**
>
> The above field is said to have an I-north-pole where the I-neg - space is in surplus.
>
> **Definition 10.2.**
>
> The above field is said to have an I-south-pole where the I-pos-space is in surplus.

Note the similarities and differences of definition 6 and 10. In both, an I-electric-field and an I-magnetic-field, the density gradient fields are formed similarly in the two parts of I-space by the compression of pockets which are "nearly" locked in their 1-1 pattern. The difference being that the compressions are due to the surplus amounts of I-neg-mass and I-pos-mass for an I-electric-field, but are due to the surplus amounts of I-pos-space and I-neg-space for an I-magnetic-field. An insulator can be used to maintain surplus amounts of I-neg-mass in a given region, away from surplus amounts of I-pos-mass at a neighboring region. It is apparently harder to find a sort-of-box to hold surplus I-neg-space in a given region away from surplus I-pos-space in a nearby region. Consequently, in an I-magnetic-field, there is always a flux of the two types of pockets being nudged in a direction opposite to that of their respective density gradient field. An I-magnetic-field can only be sustained by continually replacing the I-neg-space at its I-north pole and replacing the I-pos-space at its I-south pole. The fact that I-pos-pockets and I-neg-pockets come paired means that if there is a surplus of I-pos-pockets at one point, then there must be a surplus of I-neg-pockets at some other point. In short, an I-north-pole and an I-south-pole must come paired unless the two movements of I-pockets are completely circular. When the folding and unfolding of I-neg-space and I-pos-space is maintained in a circle, with I-neg-space rotating in one direction and I-pos-space in the opposite direction, then no I-poles are required. This case of an I-circular-magnet is discussed later. Also, there is another difference between electric and magnetic fields. Because of the nudging (flux), an I-magnetic-field tends to linger longer than an I-electric-field after the producing influence is removed. One should note that I-neg-mass causes an I-pos-electric field to form in I-pos-space, and I-pos-pockets cause an I-pos-magnetic field to form in I-pos-space.

Definition 11.

A hole in I-neg-space as given in axiom 4 is called an I-neg-gauge-boson, a hole in I-pos-space is called an I-pos-gauge-boson and a common hole, or two identical, side-by-side holes, in both parts of I-neutral-space is called an I-gauge-boson.

Definition 12.

The density gradient field formed, according to axiom 3, part 6, inside an I-gauge-boson is a reverse density gradient field (increases going out from the center of an I-gauge-boson) and is called reverse-I-gravity or I-anti-gravity.

Axiom 5. Existence and description of I-spin-orbits

There exist I-neg-spin-orbits (undefined, but are doughnuts composed of nuggets of I-neg-mass) and I-pos-spin-orbits (undefined, but are doughnuts composed of nuggets of I-pos-mass) satisfying the following:

1) The two types of spin orbits come in pairs and are produced in a spin preserving way. The number of nuggets of I-neg-mass in an I-neg-spin-orbit is equal to the number of nuggets of I-pos-mass in its paired I-pos-spin-orbit.

2) The number of nuggets used to form any I-spin-orbit is an integer and satisfies a commensurate condition (undefined, but means there are stable or "quite" stable I-spin-orbits coming in various sizes and spectrums, but not just any number of nuggets will produce a stable I-spin-orbit).

3) An I-neg-spin-orbit or an I-pos-spin-orbit possesses a plane that is determined by the maximum circle that is fully contained in the particular I-spin-orbit (doughnut), and it possesses an axis which passes through the center of this circle perpendicularly to the above plane.

4) There exists a unique point $\hat{p}$ on the maximum circle of an I-spin-orbit where unfolding and folding begin and end, and at any point p on a given I-spin-orbit (doughnut) the cross sectional area is constant.

5) Any I-spin-orbit can unfold and fold in one of two different modes.

a) The type I mode (or strong magnetic mode) is described as follows.

The unfolding takes place in a shaped chamber on the inside of an I-neg-spin-orbit producing a shaped, quantized hole of empty space that is housed in I-neg-space. The unfolding propagates in I-neg-space in a direction parallel to the axis of the given I-neg-spin-orbit. This unfolding forms an I-neg-gauge-boson. See figure 1.

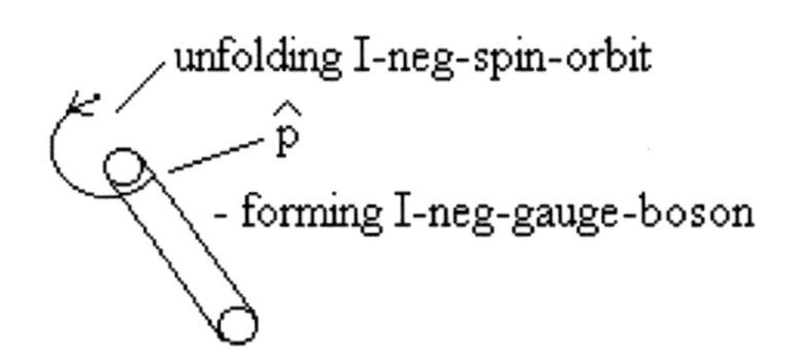

Figure 1. Type I unfolding

An I-pos-spin-orbit similarly consumes I-pos-space and forms an I-pos-gauge-boson.

b) The type II mode (or weak magnetic mode) is described as follows:

The unfolding of I-neg-space takes place on the outside of an I-neg-spin-orbit in a shaped chamber. The unfolding propagates in I-neg-space in a direction nearly parallel to the tangent line to the I-neg-spin-orbit at the unique point, $\hat{p}$. (There are fewer nuggets along the inside of an I-neg-spin-orbit than there are on its outside. Thus, the forming I-neg-gauge-boson is shorter on one side than the other, and therefore curved. This is observed in a cloud chamber.) See figure 2.

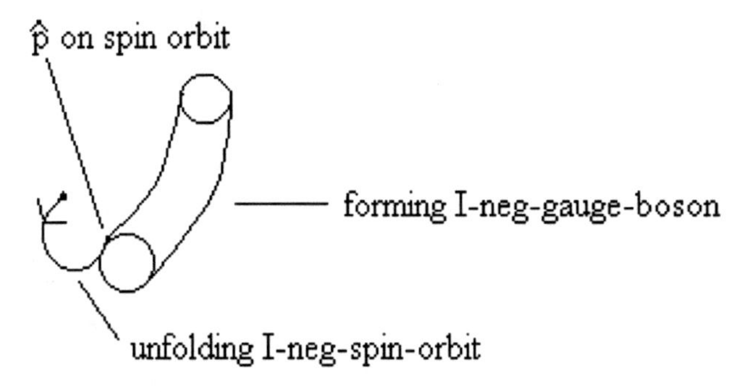

Figure 2. Type II unfolding

An I-pos-spin-orbit similarly consumes I-pos-space and forms a curved I-pos-gauge-boson.

6) An I-neg-spin-orbit with a counter-clockwise rotation, as viewed from above, produces I-neg-space "slightly" below the I-spin-orbit when in a folding phase and consumes I-neg-space from slightly above when in an unfolding phase. See figure 1. The unfolding phase pulls an I-spin-orbit in the direction of the unfolding and the folding phase pushes the forming I-neg-spin-orbit in this same direction. That is, in almost the direction of the tangent line for a type II I-spin-orbit and in the direction parallel to the axis for a type I I-spin-orbit. See figure 3.

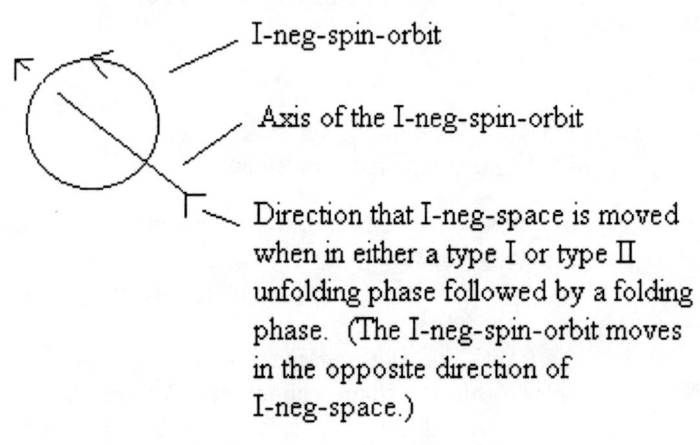

Figure 3. I-neg-space movement

An I-pos-spin-orbit with a counterclockwise rotation, as viewed from above, produces I-pos-space "slightly" below the I-spin-orbit when in a folding phase, and consumes I-pos-space from slightly above when in an unfolding phase. The unfolding phase pulls an I-pos-spin-orbit in the direction of the unfolding in I-pos-space. Also, the folding phase pushes the forming I-pos-spin-orbit in this same direction. In all cases, the more dense the pockets are, the greater the push and pull are. They are directly proportional. The times for folding and unfolding are equal.

7) The unfolding orients an I-pos-spin-orbit so that I-pos-space is consumed in the direction of the density gradient in I-pos-space. Also, the unfolding orients an I-neg-spin-orbit so that I-neg-space is consumed in the direction of the density gradient in I-neg-space. (An I-gauge-boson,

formed at an angle to the density gradient of the pockets, will be curved in the direction of the gradient. This is caused by the fact that the pockets are more dense, and therefore, shorter (smaller) on one side of the forming I-pos-gauge-boson. The next I-spin-orbit that is formed will be rotated proportionately to the amount of curving of the previous I-pos-gauge-boson.)

8) At the end of the unfolding phase of an I-spin-orbit in the folding chamber at the leading end of the associated I-gauge-boson, the angular momentum of the I-spin-orbit and the associated I-anti-gravity of the hole trigger instantaneously the starting of the folding phase. The folding continues until respectively the I-neg-gauge-boson, I-pos-gauge-boson or the I-gauge-boson formed in the unfolding phase is filled (i.e. until there is the same density inside the I-gauge-boson as there is in the surrounding I-space at the leading end of the I-pos-gauge-boson where the folding chamber exists.) There exists a critical value (not defined) for the density of pockets surrounding the leading end of an I-gauge-boson which, when exceeded, causes the folding phase to fail to take place or causes the I-gage-boson to switch directions. (The magnitude of push required to form a pocket into the very dense space cannot be produced.) Also, the folding will not take place when the energy of the I-spin-orbit goes below a certain critical value (not defined).

9) In a region of I-space, an I-neg-spin-orbit may share I-neg-space with another I-neg-spin-orbit provided they are at least slightly out of phase. (That is, one will be consuming some of the I-neg-space that the second one is currently producing and vice versa.) A similar statement holds for I-pos-spin-orbits and I-pos-space.

Definition 13.

An I-neg-spin-orbit having a rest I-neg-mass approximately equal to 9.11×10^{-28} gms is called an I-electron and its associated I-anti-particle is called an I-positron. In the type II mode of unfolding, an I-neg-gauge-boson of an unfolded I-electron is called an I-neg-photon, and an I-pos-gauge-boson of an unfolded I-positron is called an I-pos-photon. An existing I-neg-photon and an adjacent matched I-pos-photon are called an I-photon. (See axiom 6 below.)

Definition 14.

A stranded I-neg-gauge-boson, I-pos-gauge-boson or I-gauge-boson (or I-neg-photon, I-pos-photon or I-photon) is called I-neg-heat, I-pos-heat or I-heat respectively. (Stranded means the folding-unfolding chamber is destroyed and there is not a sufficiently strong triggering mechanism to start the refolding process.)

Theorem 1.

a) I-neg-spin-orbits attract I-pos-spin-orbits and vice versa.

b) An I-pos-spin-orbit repels an I-pos-spin-orbit when in a non-sharing folding and unfolding situation, and they attract one another in a sharing situation as given in axiom 5, part 9.

c) Similarly, an I-neg-spin-orbit repels an I-neg-spin-orbit when in a non-sharing folding and unfolding situation and they attract each other in a sharing situation.

Proof:

a) From axiom 3, part 7, subpart a, the I-neg-mass of an I-neg-spin-orbit causes a density gradient field of the one over distance squared type in I-pos-space which increases toward the I-neg-spin-orbit. Similarly, by axiom 3, part 7, subpart b the I-pos-mass of an I-pos-spin-orbit produces a density gradient field, again of the one over distance squared type, in I-neg-space which increases toward the I-pos-spin-orbit. From axiom 5, part 7, in the described density gradient fields the two given I-spin-orbits move toward one another in their unfolding phases.

b) Two I-pos-spin-orbits moving toward each other will each be consuming the I-pos-space that is between them. This causes a density gradient field to form pointing outward from a point midway between the pair. Thus, each will turn and move away. But when the two I-pos-spin-orbits are sufficiently close to each other with one unfolding as the other is folding, the one unfolding will be attracted to the one folding. When this unfolding is finished, the two change roles. See axiom 5 part 6.

c) A proof similar to part b holds.

Axiom 6. The nature of paired I-spin-orbits in the weak magnetic mode or the nature of I-electromagnetic waves and I-light

An I-neg-spin-orbit and a paired I-pos-spin-orbit can exist in a paired type II mode of unfolding and folding. This paired type II mode of unfolding and folding is described as follows.

1) The paired I-spin-orbits are on a common plane and they have a common tangent line on this plane. The unfolding takes place basically parallel to this tangent line. (The forces ⊥ to the tangent line cancel out.)

2) One of the I-spin-orbits has a clockwise rotation and the other a counterclockwise rotation. See figure 4.

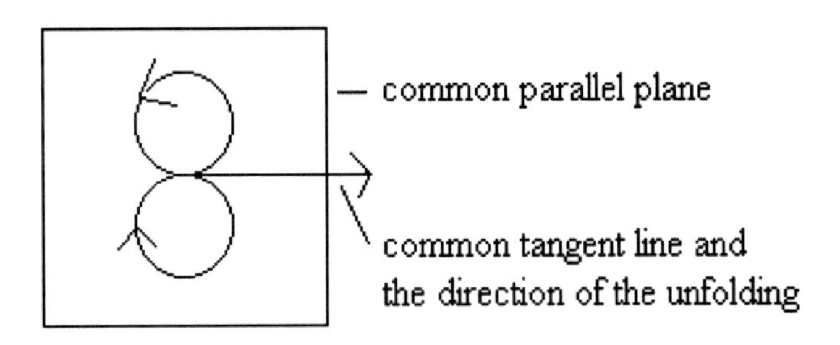

Figure 4. Paired Type II mode arrangement of unfolding

3) The common plane has no preferred orientation around the common tangent line, but once established, it remains fixed until disturbed by an outside influence.

4) At the culmination of each unfolding and at the beginning of the folding, the I-spin-momentums and the anti-gravity of the pair of I-gauge-boson holes cause the two I-spin-orbits to fold in alternating positions with their I-spin orientations reversed. (If they did not change their spin orientation, they would not be placing the I-neg-space and I-pos-space into the pair of I-gauge-boson holes properly.)

5) There is no loss of time in switching from the unfolding to the folding phase or from the folding to the unfolding phase. (If this switching is not instantaneous, then blue light will travel slower than red light. If all colors of light travel at the same speed, then this switching has to be instantaneous.)

6) The unfolding of the two I-spin-orbits forms an I-gauge-boson in I-neutral-space. The sum of the length of the two diameters of the given I-spin-orbits is "large" compared to the diameter of the cross-sectional area of each of the two sub-parts of the associated I-gauge-boson. (The cross-sectional area in uniformly dense I-neutral-space is circular with a radius proportional to the energy of the given ray. See part 11.)

7) The movement of I-neg-space and I-pos-space, as stated in axiom 5, forms an I-north pole and an I-south-pole just behind the I-spin-orbit's unfolding chamber and on a line perpendicular to the common plane of the I-spin-orbits. The two poles are on opposite sides of the I-spin-orbits' common plane. See figure 5a.

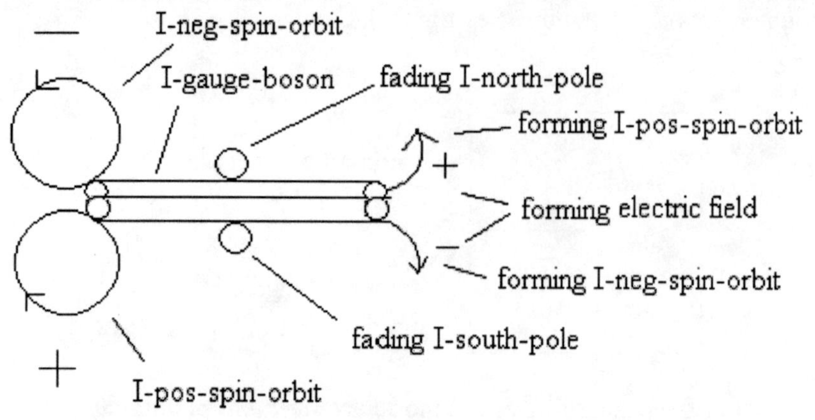

Figure 5a. I-magnetic field fades and I-electric field forms as the two I-spin-orbits are folded. The I-spin-orbits to the left disappear when the I-gauge-boson is formed. The I-gauge-boson disappears as the new I-spin-orbits are folded.

8) The unfolding and folding process described above continues until the forming I-spin-orbits are reflected, captured, or destroyed by some interference. When reflected, the unfolding and folding process continues, but it is modified in one of the following two ways.

a) Let an I-gauge-boson start forming "near" the edge of a given medium with a relatively low density of pockets of I-space. Let the unfolding end in an adjacent medium which has distinctly more dense pockets. (Note: from axiom 5, part 6, an I-spin-orbit can be formed with less push in the medium having the less dense pockets.) For reasons of ease, the folding is triggered to take place at the opposite end of the given I-gauge-boson (i.e. when compared to the normal propagation). The two I-spin-orbits fold in alternating positions, but they do not change their I-spin-orientations. See figure 5b. In this process the velocity of propagation is preserved in the direction parallel to the surface between the mediums, but the propagation velocity perpendicular to the surface is reversed. (The I-gauge-boson is filled properly.)

b) Let a medium with a relatively high density of pockets of I-space have a common surface with a medium with relatively low density of pockets. An I-gauge-boson being formed very "near" the back surface of the high density medium will unfold back into the medium having the denser pockets. (Again from axiom 5 part 6, the strong density gradient at the surface triggered this change in the direction of the unfolding.) During the next folding, the two I-spin-orbits alternate their positions, but they do not change their I-spin-orientations. See figure 5b. In this process, the velocities are affected the same as in part (a) above. (Again the I-gauge-boson is filled properly.)

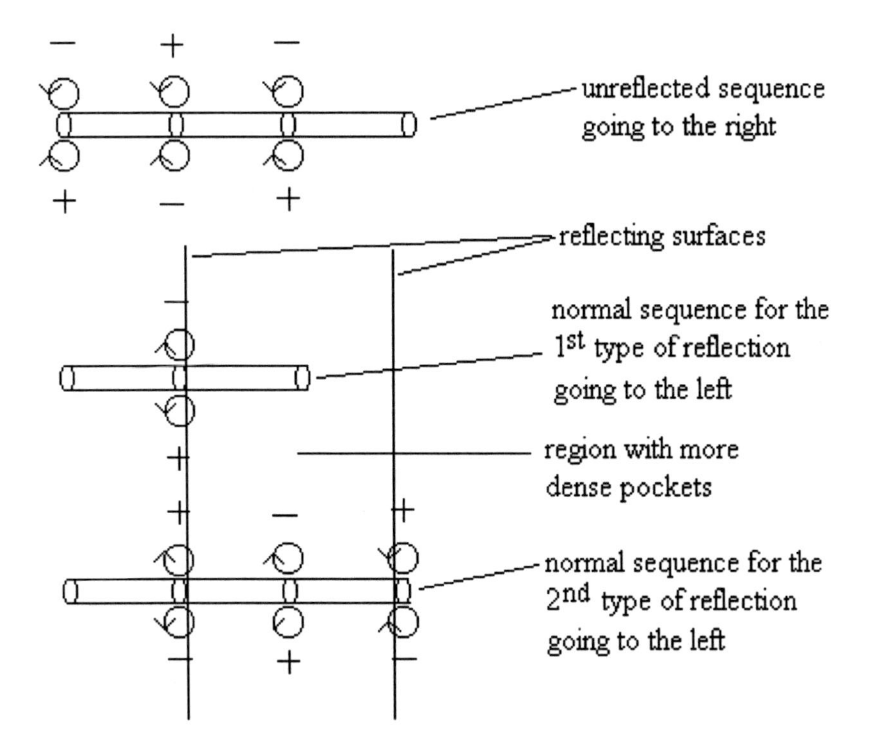

Figure 5b. Comparison of unreflected and reflected sequences

Remark: Just as there are two types of reflection, there are also, two fundamental types of I-spin-orbit interference. From axiom 5, part 9, an I-spin-orbit unfolding can consume the I-space being folded by a second I-spin-orbit of the same type. This will result in a partial or total destruction of the I-gauge-boson which the first unfolding I-spin-orbit should be forming. The destruction of all, or part, of this I-gauge-boson causes its associated I-spin-orbit to lose energy or to be destroyed in its next folding phase. If the I-spin-orbit is destroyed, what remains of the I-gauge-boson is just heat. Furthermore, there is the second, folding I-spin-orbit that has its I-gage-boson entangled with the first, unfolding I-spin-orbit's I-gage-boson. This is the second type of interference. This folding I-spin-orbit gains energy as it fills part or all of the additional, entangled I-gage boson, and it may or may not be destroyed during its next unfolding. These two types of interference show how light can have its color changed, and how the destruction of folding-unfolding chambers can produce heat and an associated increase in entropy. (In all cases the total energy is conserved by the folding-unfolding activity.)

9) Properly excited I-neutral-mass will ordinarily produce many pairs of I-spin-orbits which are very much alike and which follow each other in their unfolding and folding movements.

10) The linear velocity of the unfolding as measured in uniform I-space using a rod and clock at rest in zero gravity I-space is given by

$$\frac{d\ell}{dt} = 2c$$

where ℓ is the length of the I-gauge-boson being formed as a function of time, t, and c is the special constant approximately equal to 3.0×10^{10} cm/sec.

11) The amount of I-neutral-mass, E, unfolding with respect to distance ℓ, satisfies, for a given starting energy E_q,

$$\frac{dE(\ell)}{d\ell} = -\frac{2}{ch}E_q^2, \qquad E(0) = E_q$$

Half of the amount of unfolding comes from the I-neg-mass and the other half from the I-pos-mass.

Definition 15.

A paired I-electron and I-positron sequentially unfolding and folding, forming and erasing I-gauge-bosons along with the associated, pulsating I-electric-fields and I-magnetic-fields, as given in axiom 6, is an I-electromagnetic-wave or just an I-wave. In some cases, it is also called an I-light-wave or just I-light. See figure 6. In the case of I-light the I-gauge-boson is also called an I-photon. (See definition 13 above.)

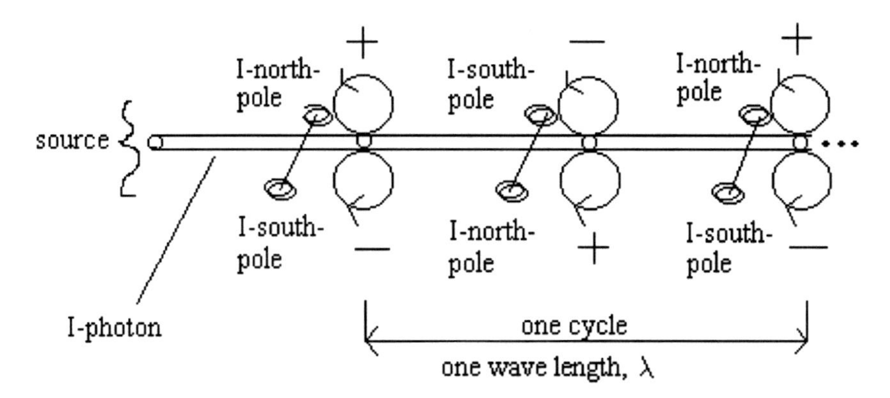

Figure 6. An I-light-wave with order of events taking place from left to right

Definition 16.

The first unfolding process in definition 15 is called "the first half of phase I" of the I-wave. The first folding process in definition 15 is called "the second half of phase I" of the I-wave. The second unfolding process of definition 15, where the I-spin-orbits are now interchanged, is called "the first half of phase II" of the I-wave. The second folding process of definition 15 is called "the second half of phase II" of the I-wave.

Definition 17.

Phase I combined with phase II of definition 16 constitute one cycle of an I-wave.

Definition 18.

The length of the two I-photons in one cycle of an I-wave is called the I-wave length and is designated by the symbol λ.

Definition 19.

A sequence of I-waves having the same I-wave-length and following one behind another is called a train of I-waves.

Definition 20.

A synchronized set of I-waves with the same I-wave length and traveling together, side by side, in the same direction, is called an I-laser-wave or I-laser-beam.

Theorem 2.

In I-neutral-space, I-light has the constant average velocity c when measured in I-space over any number of complete cycles provided a stationary rod and clock in zero gravity are used.

Proof:

From axiom 6, part 10, the velocity of the unfolding is given by

$$\frac{d\ell}{dt} = 2c.$$

The time for folding, during which there is no movement, is equal to the time for unfolding from axiom 5, part 6. There is no loss of time going from unfolding to folding from axiom 6, part 5. Thus, the average velocity for phase I is c. The same argument holds for phase II.

Remark: Pockets of I-neutral-space and nuggets of I-neutral-mass in a local region of I-space compress and expand proportionately from axiom 3, part 10. Thus, even though the length of an I-photon in dense I-neutral-space will be shorter as viewed from a housing Euclidean space, the velocity as measured in I-space will not change. This is due to the fact that the nuggets, and consequently rulers made from nuggets, will also be shorter by the same proportion as the adjacent pockets. It is very important to notice that a constant velocity for I-light moving through I-space does not mean a constant velocity for I-light moving in a housing Euclidean space. The constancy of the velocity of light in I-space is different from the constancy of light in relativity, but for "large" bodies it turns out that they are nearly the same.

Theorem 3.

For a given quantized energy, $E_q > 0$, the I-wave length λ satisfies

$$\lambda = \frac{ch}{E_q}.$$

Proof:

From axiom 6

$$\frac{dE(\ell)}{d\ell} = -\frac{2}{ch}E_q^2, \qquad E(0) = E_q.$$

Solving the differential equation gives

$$E(\ell) = -\frac{2}{ch}E_q^2\ell + k.$$

Setting $\ell = 0$, gives $k = E_q$.

Setting $E(\ell) = 0$, gives

$$\ell = \frac{chE_q}{2E_q^2} = \frac{ch}{2E_q}.$$

This states that the length of the I-photon is $\frac{ch}{2E_q}$. In turn, this is equal to $\frac{\lambda}{2}$.

Thus,

$$\frac{\lambda}{2} = \frac{ch}{2E_q}$$

or

$$\lambda = \frac{ch}{E_q}.$$

Definition 20.

The frequency, ν, for an I-wave is given by

$$\nu = \frac{c}{\lambda}.$$

Theorem 4.

The amount of energy unfolding per unit of time in an I-wave having quantized energy E_q is given by

$$\frac{dE}{dt} = -(2\nu)^2 h.$$

Proof:

$$\frac{dE}{dt} = \frac{dE}{d\ell} \cdot \frac{d\ell}{dt} = -\frac{2}{ch} E_q^2 \cdot 2c$$

$$= -4\frac{E_q^2}{h}$$

$$= -\frac{4}{h} \cdot \frac{c^2h^2}{\lambda^2}$$

$$= -4\nu^2 h$$

$$= -(2\nu)^2 h$$

Remark: The frequency for unfolding or refolding is 2 times that of the frequency for a cycle which is ν. Thus, the rate at which energy is unfolded is the frequency of half an I-wave length squared times Planck's constant. Also, this can be obtained by noticing that (2ν) nuggets of energy $\frac{E_h}{2}$ are unfolded in time $\frac{\lambda/2}{2c}$. This gives the rate to be

$$-\frac{\nu h}{\frac{\lambda/2}{2c}} = -\frac{4c}{\lambda}\nu h = -4\nu^2 h = -(2\nu)^2 h.$$

Theorem 5.

The quantized energy E_q of an I-electro-magnetic wave satisfies

$$E_q = \nu h.$$

Proof:

From theorem 3,

$$\lambda = \frac{ch}{E_q}.$$

But,

$$\nu = \frac{c}{\lambda}$$

Thus,

$$E_q = \frac{ch}{\lambda} = \nu h.$$

Remark: The number of nuggets in the quantized energy E_q is

$$\frac{E_q}{E_h/2} = \frac{2h\nu}{E_h}$$

Except for units $E_h \equiv h$, thus,

$$\frac{E_q}{E_h} \equiv \nu \text{ (in magnitude with no units)}$$

i.e. there are 2ν nuggets contained in the energy E_q. Half are I-neg-nuggets and half are I-pos-nuggets. Therefore, 2ν pockets of I-space are also involved. Note that 2ν is twice the frequency of the given electromagnetic wave.

Theorem 6.

I-light can be polarized.

Proof:

From axiom 6, part 6, the sum of the lengths of the diameters of the two I-spin-orbits is large compared to the diameter of the associated I-photon. If the plane parallel to the I-spin-orbits is also nearly parallel to the grid lines of the polarizing object, then the I-light-wave will not be destroyed. However, if the plane parallel to the I-spin-orbits is nearly perpendicular to the grid lines and the grid lines are sufficiently close together, the I-wave will be captured. Though the plane parallel to the I-spin-orbits has no preferred orientation, once established from axiom 6, part 3, it remains fixed until disturbed by an outside influence.

Theorem 7.

I-light possesses a Doppler effect.

Proof:

Let a "large" I-mass object be an I-light source and let this object be at rest in I-space. The I-photon associated with any light wave leaving this object is completely filled when the density of pockets inside the I-photon at the leading end is equal to the density of pockets that surround this leading end. Because of the I-gravitational field associated with this "large" I-mass object, the I-photon is filled back in such a way that the density inside the I-photon is not quite equal to the density of the surrounding pockets at the trailing end. Thus, more pockets were unfolded than are folded back and consequently, the wave under discussion has lost energy. This is repeated over and over. The larger the object, and the closer the wave is to this given object, the greater the loss per vibration. When the "large" I-mass object is moving in the same direction as the I-light-wave, the gravitational field is moving in this same direction and the density is slower to drop off than when the large object is at rest in the ether. This means that the energy loss, when compared to that of an at-rest "large" object, is less. Similarly, when the "large" object is moving in the opposite direction to that of the I-light-wave, then the energy loss is increased. Note that there is a loss of energy even when the "large" I-mass object is at rest in I-space, but this loss is smaller when the object is moving in the same direction as that of the I-light-wave, and this loss is greater when the object is moving in the opposite direction. This shows that a red shift in I-light indicates the source of the light is moving away. However, it should be pointed out that the amount of mass associated with the source and the amount of mass associated with the observer influences the Doppler effect. A light ray entering a gravitational field experiences an opposite change in energy compared to a light wave leaving a gravitational field. The strength of both gravitational fields enters into a light ray's observed change in energy. There is a red shift for light rays leaving a "large" mass object and being observed on a "small" mass object even when both objects are at rest in the ether.

It should be noticed that I-mass is always at rest in ether space in its I-mass state and when I-light is radiated it is always from its I-mass state. Thus, the moving source for the I-light has no influence on the energy of the radiated I-light. It is the associated I-gravitational-fields that produce the Doppler effect.

Theorem 8.

An I-light-wave will be bent in an I-gravitational-field.

Proof:

For simplicity, assume the I-light-wave is traveling at right angles to the I-gravitational-field. A forming I-photon will have denser pockets of I-neutral-space on one side than on the opposite side. By axiom 4, an equal number of pockets are unfolded regardless of the density. Thus, the edge of the I-photon formed from the denser pockets will be slightly shorter than the opposite side. Consequently, the sequence of folding and unfolding turns the folding and unfolding chamber and the I-light-wave in the direction of the density gradient of the I-gravitational-field. See figure 7.

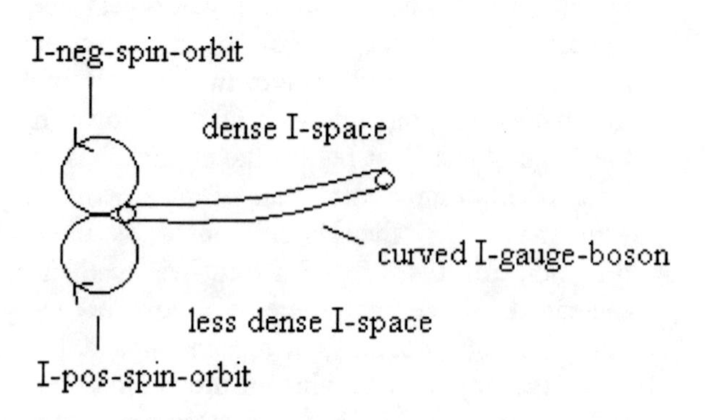

Figure 7. Curvature of I-light

Theorem 9.

I-light has the characteristics of both an I-wave and I-mass.

Proof:

At two points of time in each cycle the I-light is entirely I-mass. At two points of time in each cycle the I-light is entirely an I-photon which is void of any I-mass and is entirely an I-wave.

Theorem 10.

An I-light-wave going from any medium having one density of pockets of I-space to a second medium having a different density of pockets of I-space will be bent. (An I-light-wave will bend when entering water from the air.) The higher is the frequency of an I-light-wave, the greater is the bending.

Proof:

Assume medium 1 has less dense pockets of I-space than medium 2, and that I-light is going from medium 1 into medium 2. See figure 8.

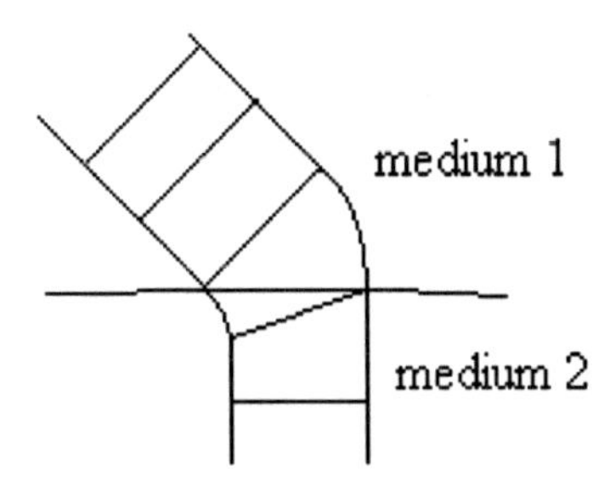

Figure 8. Bending of I-light going from one medium to another

An I-photon forming partly in medium 1 and partly in medium 2 or forming in the different densities of the two mediums, will be shorter on the side which first entered medium 2. Again, the number of pockets involved does not change because of axiom 4, and the frequency does not change, again by axiom 4.

The radius of an I-photon's cross-sectional area is proportional to the energy of its associated I-wave. The bending starts when a point of a forming I-photon contacts the surface and stops when the leading cross-sectional area of the folding and unfolding chamber of the I-photon is entirely in the new medium. The rate of bending is determined by the difference between the velocities of the I-wave in the two mediums. The larger the diameter of the

I-photon, the longer is the time of turning and the greater is the bending. High frequency I-light-waves have larger diameters than low frequency I-light-waves and consequently, bend more.

Theorem 11.

I-light in a medium having less dense pockets of I-space will have a greater velocity than I-light in a medium having denser pockets of I-space when viewed from a housing Euclidean space.

Proof:

For a given fixed quantized energy, E_q, axiom 4 states that the number of pockets unfolded to form an associated I-photon is also fixed and is independent of the density of the pockets. When the length of the I-photon is viewed in a housing Euclidean space it is shorter in the I-space with the more dense pockets. The frequency is independent of density and consequently, is the same in the two mediums (as long as the clocks used have the same velocity with respect to I-space. See theorem 16, part b.) This gives the desired result.

Theorem 12.

I-light possesses momentum.

Proof:

In the folding phase of I-light, there are two I-spin-orbits forming and an I-photon being filled with I-space as shown in figure 9.

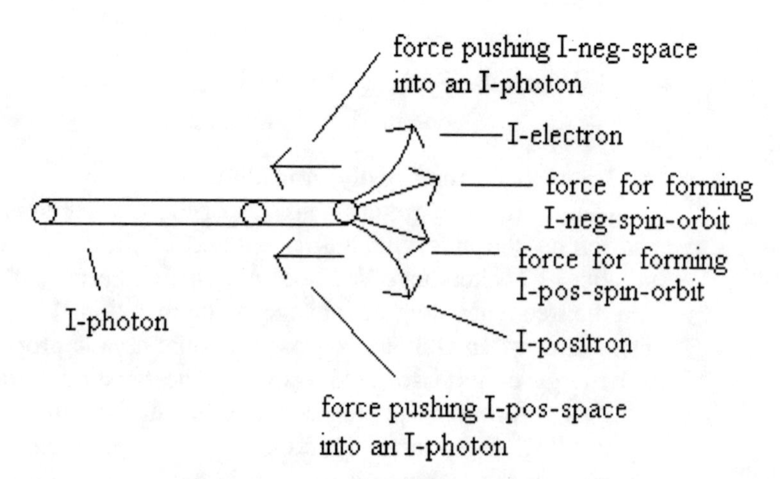

Figure 9. Momentum of I-light

When an I-photon is being filled, there is a forward force on the forming I-spin-orbits and a backward force on I-space from axiom 5, part 6. As always, the sum of all the forces is zero, because to every action there is a reaction. The vertical forces components cancel as do the horizontal forces. If the two spin orbits are being captured, then the forward force is exerted on the capturing I-particle. This force is the momentum force. (Note that the I-spin-orbits form circularly because they cannot pass through I-space, they can only compress the I-space pockets. This compression produces an I-electric-field when the I-spin-orbits are formed.)

Theorem 13.

An intense beam of I-light will diffuse when passing through a small opening.

Proof:

A beam of I-light causes a decrease in the average density of I-space within the beam. This is due to the large number of I-photons forming and being filled and, therefore, yielding a smaller average density of I-space on the interior of the beam than exists on the outside of the beam. Thus, there is an anti-gravitational field perpendicular to the beam's line of motion. By theorem 8, this causes I-light on the edge of the beam to be bent away from the center of the beam of I-light. Thus, I-light is diffused.

Theorem 14.

I-light can interfere with other I-light and cause a change in the frequency and wavelength, i.e. a change in color.

Proof:

If two I-photons become entangled head to tail and two or more spin orbits are formed side by side from these two I-photons, the wavelength will be made approximately twice as long, and the frequency will be cut in half.

Theorem 15.

An I-laser-beam or any I-heat increases the energy of I-mass folding into the associated hole.

Proof:

An I-laser-beam of I-light has a large number of I-photons adjacent to each other forming in essence one larger hole. Except for the fact that this larger hole of I-photons is traveling as I-light, it acts exactly the same as I-heat. Any I-mass folding and coming in contact with this hole will increase its energy by folding additional I-mass.

Axiom 7. The nature of paired I-spin-orbits in the strong magnetic mode or the nature of velocity's influence on I-mass, time and length.

An I-neg-spin-orbit and a paired I-pos-spin-orbit come formed together and possess a paired type I mode of unfolding and folding. See figure 1. The properties of this paired, type I mode of unfolding and folding are delineated in the following statements:

1) Two closed, paired I-spin-orbits coexist in 3-space and have a common spin-axis through their common center. The plane that is perpendicular to this common axis at the common center point is called the plane of symmetry. See figure 10 and figure 11.

2) One I-spin-orbit has a clockwise rotation and the other has a counterclockwise rotation, and they unfold and fold in sync. See figure 10.

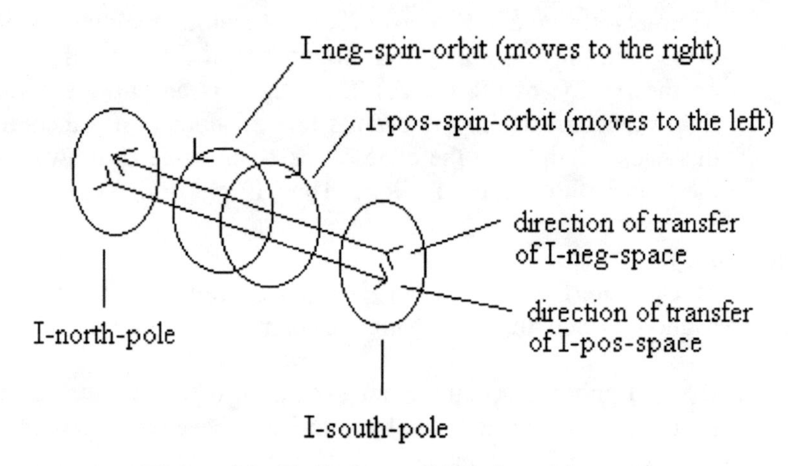

Figure 10. Paired type I I-spin-orbits

3) The plane of symmetry has no preferred orientation, but once established, remains invariant until disturbed by a density gradient field in the pockets of I-space. The orientation process referred to in axiom 5, part 7 applies to each of the paired I-spin-orbits. This means that the unfolding in a density gradient field pulls on the pair of I-spin-orbits making them have a common diameter which is parallel to the local density gradients. Next this same pull causes each of the paired I-spin-orbits to pivot open. This pivot rotation is made around their common diameter that is perpendicular to both of their axes and to the local density gradients. This

means that the velocity vector ends up being parallel to the local density gradients. See figure 11.

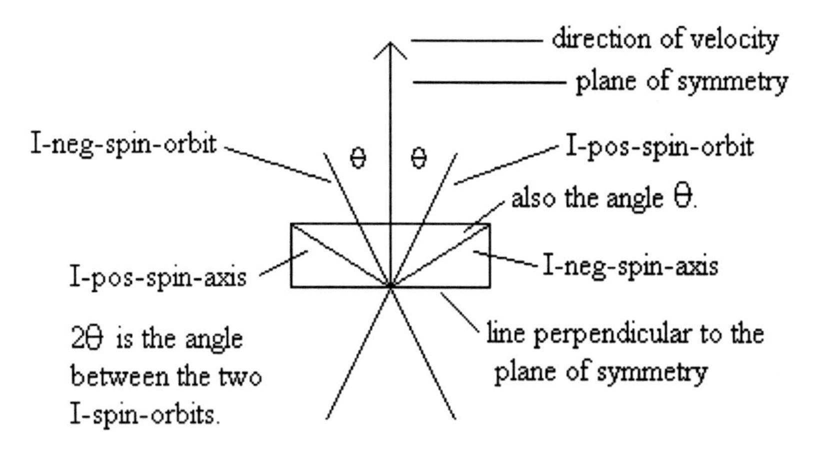

Figure 11. Paired type I I-spin-orbits

(The density gradient field that causes this orientation can be that of I-gravity or that of a density gradient field formed on the front side of any body of I-mass being forced into I-space. For accelerated motion, or for just plain motion, the more dense pockets form ahead, in the direction of the force vector and/or the direction of the velocity vector. This leaves less dense pockets behind. A combination of all of these density gradient vectors is obtained by a vector summation. Thus, I-gravity and/or any other symmetric in the two parts of I-space density gradient field cause paired I-spin-orbits to open or spread-out in a symmetrical fashion.)

4) At the culmination of each unfolding, the I-spin-momentums cause the two I-spin-orbits to fold again, keeping their same relative positions. (They do not alternate positions as in an I-electromagnetic-wave.)

5) There is no loss of time in switching from the unfolding phase to the folding phase, or from the folding phase to the unfolding phase, in this paired mode of unfolding.

6) The unfolding of the I-neg-spin-orbit forms an I-neg-gauge-boson and the unfolding of the I-pos-spin-orbit forms an I-pos-gauge-boson, just as in figure 1 of axiom 5.

7) The movement of I-neg-space and I-pos-space, as stated in axiom 5, still holds, i.e. an I-north-pole and an I-south-pole form along the axes on the outside of the paired I-spin-orbits in a way consistent with axiom 5. See figure 10.

8) The unfolding and folding continues until the I-spin-orbits are destroyed or broken up. Such a break-up takes place in an atomic bomb.

9) The linear velocity of the unfolding along the axis of an I-spin-orbit as measured in I-space is given by

$$\frac{d\ell}{dt} = 2c.$$

This is just the same as in the type II mode. See axiom 6, part 10.

10) The amount of I-neg-mass or I-pos-mass, $E(\ell)$, unfolding with respect to distance, ℓ, satisfies for a given, starting at rest, energy, E_0,

$$\frac{dE(\ell)}{d\ell} = -\frac{1}{ch}(2E_0)^2, E(0) = E_0$$

Note that this equation is just half of the equation given in axiom 6, part 11. This is because the paired I-spin-orbits unfold in opposite directions. Thus, there is either only an I-neg-spin-orbit or only an I-pos-spin-orbit under consideration in calculations involving the rate of linear unfolding. Also, instead of having an equal number of I-pos-nuggets and I-neg-nuggets, here one has two times as many I-pos-nuggets or two times as many I-neg-nuggets inside the square term.

11) For each phase of unfolding, no matter how the paired spin-orbits are oriented with respect to each other, for any given pair of I-spin-orbits the cumulative length of unfolding perpendicular to the plane of symmetry is always the same distance and involves the same number of pockets of each type. This distance is equivalent to the cumulative linear length of the pockets unfolded when the paired I-spin-orbits are closed and operating alone. The fact that there is shared folding and unfolding in I-swirls means that the I-gauge-boson produced is usually a virtual I-gage-boson because it is forming and being filled at the same time. This is why the word "cumulative" is needed when describing the distance covered.

Definition 20.

An I-neg-spin-orbit and a paired I-pos-spin-orbit in the type I mode of unfolding and folding is called an I-gyro of the first kind. When in the type II mode of unfolding and folding, it is called an I-gyro of the second kind. These are designated I-gyro-1 and I-gyro-2, respectively. An I-gyro-1 is sometimes referred to as just an I-gyro. (The I-gyros-1 are strong on I-magnetism, have a variable velocity, but a constant cross-sectional area for unfolding. The I-gyros-2 are weak on I-magnetism, have a constant velocity, but have a cross-sectional area for unfolding that can vary.)

Definition 21.

The total angle between the planes which are adjacent respectively, to the faces of the paired I-spin-orbits of an I-gyro-1, is called the spread angle. See the angle 2θ in figure 11. (It is shown in theorem 16 below that this angle determines the velocity of an I-gyro-1, and at the I-gyro-1 level accounts for the shortening of sticks in the direction of motion. Also, this angle is associated with the increase in I-mass, and the slowing down of time for a given increase in velocity.)

Definition 22.

In I-neutral-space, the line formed by the intersection of the plane of symmetry with the plane containing non-parallel, paired axes of an open I-gyro-1, and directed from the center of the I-gyro-1 toward the unfolding side, is called the direction-of-motion. See figure 11.

Definition 23.

The force required in the production of a density gradient field sufficiently strong to make the direction-of-motion of an I-gyro-1 or of a group of I-gyros-1 parallel to a sought velocity vector is called I-initial-inertia. (It takes a little extra force to orient I-gyros-1s to prepare them for a change in velocity.)

Definition 24.

The length of the I-neg-gauge-boson or the I-pos-gauge-boson of an I-gyro-1 is called half the wavelength, $\lambda/2$, of an I-neg-spin-orbit or an I-pos-spin-orbit, respectively. ($\lambda/2$ is used here to coincide with the I-gyros-2, but here the two phases of these cycles are identical.) Recall, over any number of complete cycles, the average linear velocity of unfolding and folding is c. The number v, such that, for wavelength, λ,

$$v\lambda = c$$

is called the frequency of the I-gyro-1.

Theorem 16.

a) If m_0 represents the I-mass of an I-gyro1 at rest, and m is its I-mass when moving at a constant linear velocity, v, then

$$m = \frac{m_0}{\sqrt{1 - \frac{v^2}{c^2}}}.$$

Proof:

From axiom 7, part 9, the linear velocity of unfolding along the axes of the I-spin-orbits of an I-gyro-1 is given by

$$\frac{d\ell}{dt} = 2c.$$

From axiom 3, part 5, in order to move through I-space, there must be the proper amount of unfolding in the direction of the velocity. Let $\lambda/2$ be the cumulative length of the I-neg-gauge-boson for an at-rest I-neg-spin-orbit unfolding. For the initial at-rest energy, E_0, the differential equation from axiom 7, part 10 is

$$\frac{dE}{d\ell} = -\frac{1}{ch}(2E_0)^2, \quad E(0) = E_0 \text{ and } E(\tfrac{\lambda}{2}) = 0.$$

Solving,

$$E(\ell) = -\frac{1}{ch}(2E_0)^2 \ell + E_0.$$

Setting $\ell = \frac{\lambda}{2}$ yields,

$$0 = -\frac{1}{ch}\frac{\lambda}{2} 4E_0^2 + E_0.$$

Thus, for $E_0 > 0$,

$$E_0 = \frac{ch}{2\lambda}$$

or

$$E_0 = \nu(\tfrac{h}{2}).$$

A similar result holds for the I-pos-spin-orbit. Thus, for the whole I-gyro-1, its rest I-mass is $2E_0$ where $2E_0 = \nu h$. For movement parallel to the motion vector for a fixed, linear velocity, v, the ratio of unfolding along the axis to the unfolding in the direction of the velocity must be $\frac{2c}{2v}$ or $\frac{c}{v}$. Let θ be half of the spread angle. See figure 11. Then,

$$\sin\theta = \frac{v}{c}$$

and the anti-I-mass pockets' velocity is now $c\cos\theta$ where

$$\cos\theta = \sqrt{1-\tfrac{v^2}{c^2}} \quad \text{and} \quad m_0c^2\sin^2\theta + m_0c^2\cos^2\theta = m_0c^2 .$$

The folding phase takes the same amount of time as the unfolding phase, and takes place at the same rate as the unfolding phase. Therefore, the folding energy, $E(\ell)$, satisfies

$$\frac{dE}{d\ell} = \frac{1}{ch}(2E_0)^2, \; E(0) = 0$$

And thus

$$E = \frac{1}{ch}(2E_0)^2\,\ell .$$

The cumulative length of the component of unfolding perpendicular to the velocity vector has the constant value, $\lambda/2$. See axiom 7, part 11. Let ℓ be the length of the unfolding along the axis, i.e. along the hypotenuse of the right triangle under discussion. Then,

$$\ell = \frac{\lambda/2}{\cos\theta} = \frac{\lambda}{2\sqrt{1-\tfrac{v^2}{c^2}}} \quad .$$

Putting this ℓ in the equation for the amount of energy folded over this distance gives

$$E\left(\frac{\lambda}{2\sqrt{1-\tfrac{v^2}{c^2}}}\right) = \frac{1}{ch}(2E_0)^2\frac{\lambda}{2\sqrt{1-\tfrac{v^2}{c^2}}}$$

or

$$E\left(\frac{\lambda}{2\sqrt{1-\frac{v^2}{c^2}}}\right) = \frac{1}{ch}\frac{(ch)^2}{\lambda^2}\frac{\lambda}{2\sqrt{1-\frac{v^2}{c^2}}}$$

$$= \frac{ch}{2\lambda}\frac{1}{\sqrt{1-\frac{v^2}{c^2}}}$$

$$= E_0 \frac{1}{\sqrt{1-\frac{v^2}{c^2}}}.$$

This implies that, E_v, the energy at velocity v, is given by

$$E_v = \frac{E_0}{\sqrt{1-\frac{v^2}{c^2}}}$$

or in terms of I-mass, using $E = mc^2$,

$$m = \frac{m_0}{\sqrt{1-\frac{v^2}{c^2}}}.$$

A similar proof holds for the I-pos-spin-orbits. Thus, this result holds for the I-mass of the whole I-gyro-1.

b) The time t on the moving I-gyro-1 satisfies

$$t = \frac{t_0}{\sqrt{1-\frac{v^2}{c^2}}}$$

where t_0 is the time for a quarter of a cycle while at rest. Note that these times here are units of time. With an increase in velocity the unit of time increases which means that time slows down.

Proof:
The linear unfolding rate is the constant $2c$. Thus, at rest, the time for a quarter of a cycle is

$$t_0 = \frac{\lambda/2}{2c} = \frac{\lambda}{4c}.$$

Using the length of the unfolding along the hypotenuse computed above when moving with velocity v, the time, t, is given by

$$t = \frac{\dfrac{\lambda}{2\sqrt{1-\frac{v^2}{c^2}}}}{2c} = \frac{\lambda}{4c}\frac{1}{\sqrt{1-\frac{v^2}{c^2}}}$$

or

$$t = \frac{t_0}{\sqrt{1-\frac{v^2}{c^2}}}.$$

As the time for each quarter of a cycle satisfies this relation, the same result holds for any number of cycles. Also, all movements of any I-gyro-1 depend on the time for unfolding and folding. Consequently, this time relation holds for all movements of any I-gyro-1.

c) If D is the diameter of an I-gyro-1 at rest, then its projected length in the direction, v, is $D\sqrt{1-\frac{v^2}{c^2}}$.

Proof:

The projected length is $D\cos\theta$. See figure 11. Thus,

$$D_{proj} = D\sqrt{1-\frac{v^2}{c^2}}.$$

(The I-gyro-1 has become shorter in the direction of motion.)

d) The unfolding rate for the I-neg-spin-orbit over half an I-wave length satisfies

$$\frac{dE}{dt} = -\frac{1}{2}(2v)^2 h.$$

Proof:

$$\frac{dE}{d\ell} = -\frac{1}{ch}(2E_0)^2$$

$$\frac{dE}{dt} = \frac{dE}{d\ell} \cdot \frac{d\ell}{dt} = -\frac{1}{ch}(2E_0)^2 \cdot 2c$$

$$= -\frac{2}{h}(vh)^2 = -2v^2 h$$

$$= -\frac{1}{2}(2v)^2 h$$

A similar result holds for I-pos-spin-orbit. Therefore, for the whole I-gyro-1, over half an I-wave length

$$\frac{dE}{dt} = -(2v)^2 h.$$

Repeated Remark: For an I-gyro-1, velocity is variable, but the cross-sectional area of the unfolding chamber is invariant. For an I-gyro-2, the cross-sectional area of the unfolding chamber is variable, but its velocity is invariant. Also, an I-gyro-1, held fixed in an I-gravitational field, will open consistent with the pull of I-gravity. Therefore, any held I-gyro-1 will have the three characteristics of this theorem 16.

Axiom 8. Existence and Structure of I-quarks and I-atoms

There exist I-quarks and I-atoms possessing the following structure:

1) I-quarks and I-atoms consist of sets of I-gyros of the first kind in close proximity with adjacent I-gyro-1s unfolding and folding completely out of sync. The I-gyros-1 are attracted into I-swirls (undefined, but semi-stable I-swirls have a commensurate condition on the number of I-gyros used to form each of them. Also, each I-swirl, in the group of I-swirls forming an I-quark or an I-atom, have one or two, or possibly more, regions where the curvature of the I-swirls is much larger than on the other parts of any I-swirl. At each of these regions there is either a longer circuit of I-neg-space which produced an I-north-pole and in addition makes the associated swirl appear to be positively charged, or there is a longer circuit of I-pos-space which produces an I-south-pole and in addition makes the

associated I-swirl appear to be negative. Because of the poles and apparent charges, swirls are attracted together to form I-quarks and I-atoms. An I-swirl is "nearly" a circular magnet.)

2) The I-gyros-1 in an I-swirl share I-space, i.e. an I-gyro-1 in a folding phase will be producing I-neg-space to one side and I-pos-space to the opposite side while, at the same time, the two neighboring I-gyro-1s of the I-swirl (one on each side) are unfolding the I-space being produced.

3) The forming of an I-swirl has the following associated phenomena:

a) An I-gyro-1 being pulled into an I-swirl will not always be completely synchronized with the adjoining I-gyro-1 producing I-space, i.e. the I-gyro-1 being pulled into the I-swirl will find that the I-gyro-1 producing I-space for its unfolding will itself go into the unfolding phase before the entering I-gyro-1 has completed its unfolding. At this point both I-gyros-1 in question begin to unfold and compete for the I-space that is between them. The two unfolding I-gyros-1 cannot finish unfolding without finding additional I-space outside of the given I-swirl. This unfolding on the outside of the I-swirl produces side by side holes in I-neg-space and/or I-pos-space. The folding to fill these holes results in forming I-pos-particles and/or I-neg-particles on the outside of the given swirl (for example I-pions). The I-energy of these particles is dependent on the amount of I-space that ends up being shared in the I-swirl. The greater the sharing, the smaller the I-mass of the departing I-particles. In the special case of going from combined I-neutron-swirls to combined I-proton-swirls the emitted I-particles form a quantized I-gyro of the second kind and an associated orbiting, identically quantized I-electron with the associated identically quantized I-positron staying in the internal circuits of the swirls. When this happens an I-neutron is changed into an I-proton and this process is very hard to reverse. The total I-mass (or I-energy) is always conserved by the folding-unfolding activity. Many of the details of the way I-swirls and I-quarks are formed are left open.

b) The sharing of I-space in an I-swirl results in a surplus of I-space outside of the I-swirl.

c) Any quantized I-gyro of the second kind, composed of an I-electron and an I-positron, can become associated with an out-of-sync orbiting and identically quantized I-electron moving in the strong I-gravity-field of the given I-swirl. This makes the I-electron's orbit move further out. (See axiom 5, part 5, b.)

Note that an orbiting I-electron can possess an attached energy, provided a matching amount of energy is available to form an out-of-sync, adjoining I-neg-spin-orbit that is identically quantized. At the same time that this new adjoining I-neg-mass is folding and unfolding out of sync with the orbiting I-electron, there is an associated I-pos-mass called an I-positron that is folding and unfolding along with it in a spin preserving way. When everything is in order, there is an I-gyro of the second kind attached to the original orbiting I-electron. This hybrid type of orbiting I-electron is forced into a higher orbit. The attached I-gyro-2 can now be radiated off to form an I-electromagnetic wave. When this occurs, the I-electron drops back to a lower orbit. When it drops back to its basic orbit, it no longer can radiate because all of the attached I-gyros of the second kind are gone. This explains the characteristic absorption and radiation spectrum lines associated with each type of I-atom.

4) When a force is exerted on the I-swirls in an I-atom, the I-swirls pivot so that the plane of symmetry for every I-gyro-1, of the given I-swirl, is capable of containing the force vector. The resulting acceleration and velocity, v, will be in the direction of the force vector. See figure 12. Also, an I-atom inherits its I-mass, length and time characteristics from the I-gyros of its I-swirls. The I-north-pole and the I-south-pole can be nearly at the same point or spread out in an I-swirl.

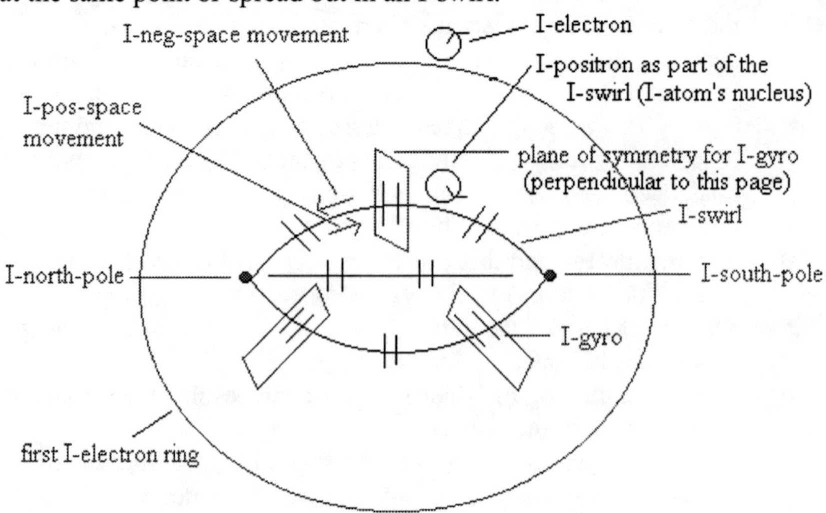

Figure 12. I-swirls in the plane of this page

Definition 25.

Neighboring I-gyros of I-atoms having a large proportion of their I-north-pole to I-south-pole orientations aligned parallel to each other over a given body of I-mass, is called an I-magnet. The complete story by which this is accomplished is left open. Repeating, the I-poles are shown here as symmetrical and a long ways apart, but they also, can exist very close to one another.

Theorem 17.

In I-space, an I-north-pole of one I-magnet is attracted to an I-south-pole of a second I-magnet, while like I-poles repel each other.

Proof:

> By reason of the symmetry stated in the axioms, an I-neg-spin-orbit and its paired I-pos-spin-orbit folding or unfolding in I-neutral-space have canceling pushing or canceling pulling taking place in the two I-spin-orbits. However, when a separate I-north-pole is held next to an I-south-pole, at the I-south-pole the density of I-neg-pockets is increased and the density of I-pos-pockets is decreased making the pull on I-neg-pockets toward the I-north-pole increase and the push away on I-pos-pockets decrease. A similar statement holds for the I-north-pole. Further, the I-south-pole is removing I-neg-pockets from between the I-poles and the I-north-pole is removing I-pos-pockets from this same region. When the rate of removing pockets of the two types exceeds the rate at which pockets are nudged into or are formed between the two I-poles, the room between the two I-poles begins to disappear. An I-north-pole and a near by I-south-pole are now attracting each other. Similarly, in the case of like I-poles, the pulls together decrease, the pushes away increase and additionally, the number of pockets of one type or the other begin to increase between the two I-poles. The two like I-poles are now repelling each other.

Axiom 9. Existence of Solid, Liquid and Gaseous States Along with Molecules.

At high energies I-atoms repel each other due to the action of the circulating electrons. This produces a gaseous state. At medium energies the I-atoms tend to lock in 2-dimensional films, I-north-pole to I-south-pole, producing a liquid state. At low energies, the I-atoms tend to strongly lock in a 3-dimensional form, I-north-pole to I-south-pole, producing the solid state. I-atoms of different types, when properly matched as determined by the number of circulating electrons of the various I-atoms, tend to strongly lock I-north-pole to I-south-pole. All details of these processes are left open.

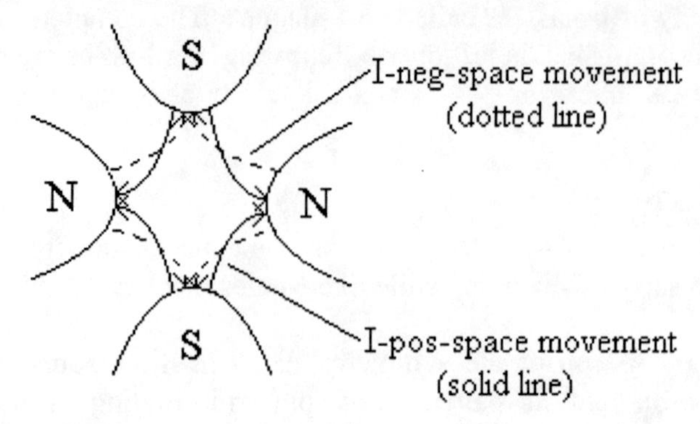

Figure 13. Liquid state, I-magnetic, two-dimensional, 4 atoms locking. 2, 6, etc. lockings would look similar.

Axiom 10. Behavior of I-electrons on the surface of a charged wire.

When a surplus of I-positrons exist on one isolated body, and a wire is attached to this body, a strong gradient in I-neg-space is set up on the surface of the wire, pointing along the wire in the direction of the excess I-positrons. Upon connecting this wire to a body with a surplus of I-electrons, the I-electrons travel unfolding and folding in the tangential mode toward the body having the surplus I-positrons. The unfolding and folding of the I-electrons along the wire moves I-neg-space from the right side of the moving ahead I-neg-spin-orbit to the left side. This creates a movement of I-neg-pockets around the wire. See figure 14.

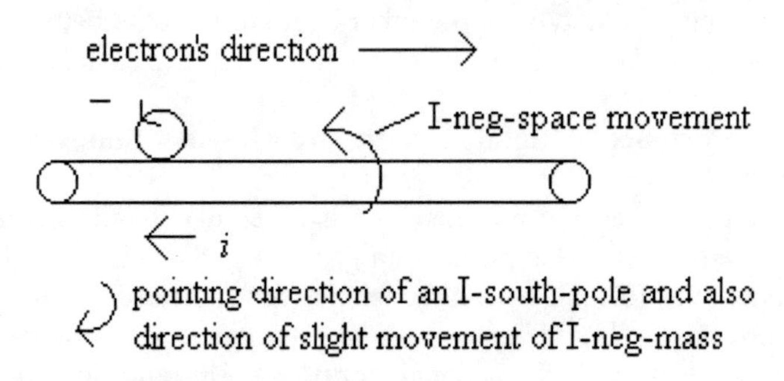

Figure 14. Electric field around a wire

Remark: For proofs on electricity, replace lines of force with "direction of movement of I-neg-space." Then the proofs are about the same as the reasoning using lines of force. For example, in the case of a wire carrying a current in an I-magnetic field, the force on the wire is produced as follows.

In I-space, let an I-magnetic field nudge I-neg-space to the right. Let I-neg-space be pumped around a wire which is positioned perpendicularly to the I-magnetic field and is carrying a current. Then on the side of the wire where the I-neg-space is pumped to the left, meeting the I-neg-space being nudged to the right by the I-magnetic field, the I-neg-pockets become compressed and therefore, more dense. See figure 15. Some of this more-dense I-neg-space pairs up with the surplus I-pos-space coming from the I-south-pole producing an I-gravitational field. On the opposite side where the I-neg-pockets are going in the same direction, the I-neg-pockets are not compressed as much and there is less pairing up of the I-neg-space and the I-pos-space. The moving electrons and the gyro-1s of the wire now pull the wire toward the more dense I-neg-pockets. (This is in contrast to arguing that the wire is pushed away from the denser lines of force which are all going in the same direction. But the two methods do give the same result.)

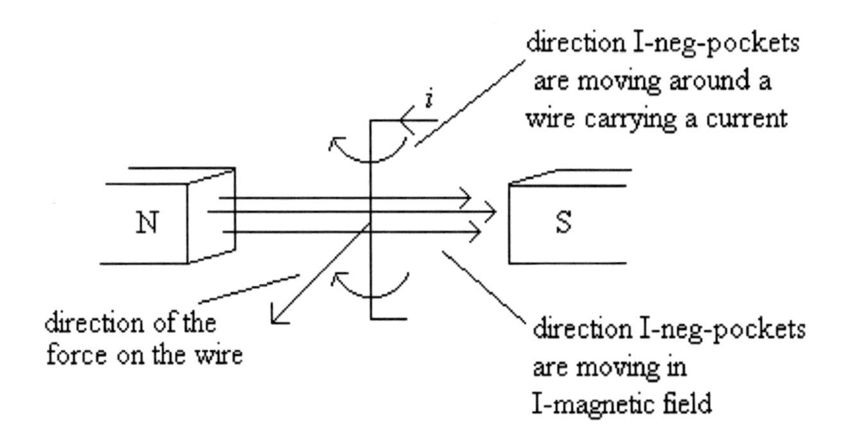

Figure 15. Dense pockets of I-neg-space being pushed together, along with the I-gyros of the wire, create a force on a wire carrying a current.

Note that there is no nudging of the I-pos-pockets around the wire carrying a current. The moving electrons, making up the current, pump only the I-neg-space.

Remark: The I-space introduced in the foregoing 10 axioms does not have an associated turmoil ether wind. If the pockets of I-pos-space and I-neg-space are in equilibrium, there is no flowing, only compressing or expanding. When the pockets are not in equilibrium, only the one in surplus is moving. The I-gravitational fields around a stationary body (earth) and a slow moving body (compared to the speed of I-light) each having the same I-mass are essentially identical. In the direction of motion, the I-gravitational field is just slightly below the one over distance-squared curve and it is building up and adjusting very rapidly. While in the opposite direction to the direction of motion, the I-gravitational field is just slightly above the one over distance-squared curve and it is dying down very rapidly. This described phenomenon produces the rendezvous effect. This rendezvous effect along with moving gravitational fields imply that Michelson and Morley's experiment cannot detect any difference between a "large" body at rest and this same body in "slow" motion relative to the speed of I-light. See theorem 24.

The reason a karate punch is so effective is that a quick impulse in the pockets of I-space yields a sharp density curve over a short distance compared to a slower-applied impulse which has a more flattened density curve. See figure 16.

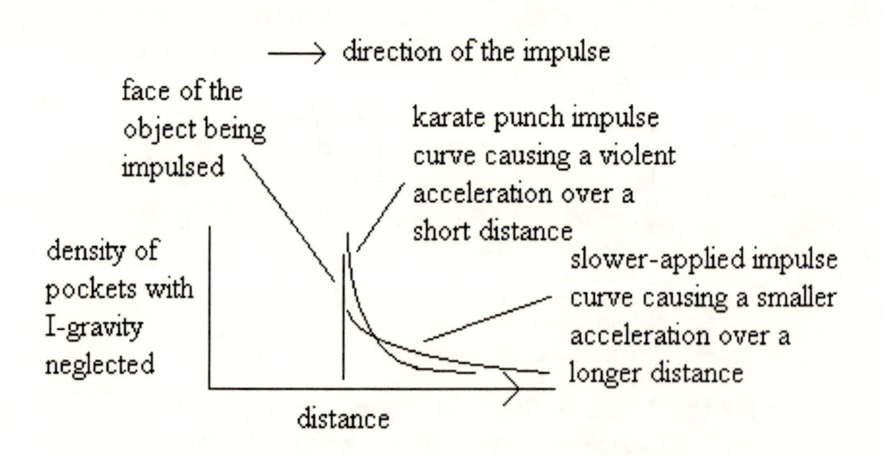

Figure 16. Karate punch vs slower-applied impulse

The movement of I-mass in I-space is just the opposite of the commonly accepted movement of mass in space. Mass is said to move uninhibited in a vacuum. While in I-space, I-mass is completely boxed in and can only move by unfolding a path for its movement. All I-mass movement is the result of reacting to the I-space adjoining it, and there is no force at a distance.

Theorem 18.

An I-light-wave traveling perpendicular to the surface of a medium with dense pockets and reflected back into a medium having less dense pockets, will have a half wave length change in phase.

Proof:

> See axiom 6, part 8 and note that an I-photon has a length equal to half a wavelength.

Theorem 19.

Very thin soap films will reflect little light, slightly thicker ones reflect violet or blue, still thicker, yellow or red. When the thickness is much greater, the reflected light is white.

Proof:

> From axiom 6, part 8, an I-gauge-boson starting to form "near" the surface of the soap bubble will be reflected when the leading end of the I-gauge-boson is still in the medium (film) having the denser pockets. Consequently, if the thickness of the film of the soap bubble is just a little greater than that wavelength of an I-light ray, then this ray can be reflected in the two ways of axiom 6, part 8. (If it is a little thinner, there is very little reflection possible.)
>
> This means that a short I-light-wave has two chances to be reflected, while a slightly longer I-light-wave can rarely be reflected. (The I-photon of the longer I-wave lengths, which starts forming near the surface of the soap bubble, will reach completely through the film. These I-light-waves have no reason to change direction.)
>
> When the thickness of the film is a little longer than the I-wave lengths of yellow or red, these I-light-waves can be reflected in two ways. At the same time, the violet or blue I-waves being reflected in two ways will form some yellow or red light. This occurs when the I-photon of the first type of reflection matches up with an I-photon of the second type of reflection. See figure 17.

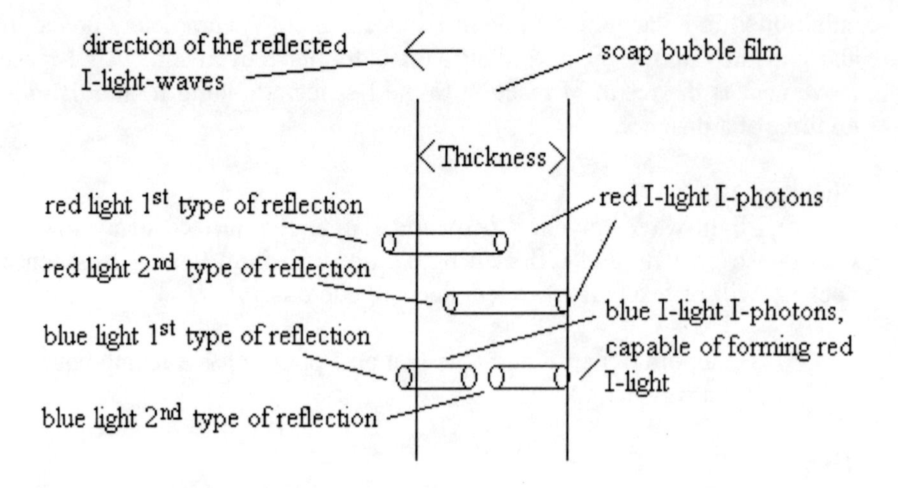

Figure 17. Reflection of red and blue I-light

When the soap bubble's thickness is large, all I-wave lengths of light have nearly an equal chance of reflecting. In this case, the reflected I-light is white.

Theorem 20.

I-mass cannot attain an absolute zero temperature at all points in its interior.

Proof:

Reducing the energy of I-mass by continually lowering its temperature leads to one of the two following possibilities.

1) At no point in the I-mass is an absolute zero temperature attained.

or

2) At some point in the I-mass an absolute zero temperature is attained.

In 2), if an absolute zero temperature is attained at some point in the I-mass, then the energy will be sufficiently low, and the I-spin-orbit will not refold. (See axiom 5, part 8.) This results in the formation of an isolated I-gauge-boson, which in turn heats up the I-mass around it, thus keeping it from attaining absolute zero. Consequently, in either case, no point in the I-mass is at absolute zero.

Theorem 21.

In I-space, the dying of a Black Hole heralds the birth of a Quasar or some similar entity.

Proof:

> At the center of a Black Hole, the nuggets of I-mass and the pockets of I-space become very compressed. From axiom 5, part 8, it becomes harder and harder for the folding process to take place. When the density of the pockets that form I-space exceed the critical value referred to in axiom 5, part 8, a large number of stranded I-gauge-bosons arise. This occurs because the folding phase cannot take place. These stranded I-gauge-bosons are I-heat. In addition, when there are a sufficient number of these stranded I-gauge-bosons, a very strong I-anti-gravity field develops in the big hole formed by these combined stranded I-gauge-bosons. This strong anti-gravity field, along with the intense I-heat, literally blows hot I-mass out from the center at the Black Hole. In I-space, this is the birth of a Quasar or something similar.

Theorem 22. (Erica's Theorem)

A beam of I-electrons flowing head on into a beam of I-positrons pass through each other with very few collisions, even though they are attracted to each other.

Proof:

> From axiom 3, I-space is composed of two intermingled, but distinct, finite sets of pockets, one set filled with I-neg-space and the other with I-pos-space. As the stream of I-positrons (I-pos-mass) flows through this I-space, it folds and unfolds with the I-pos-space, hardly affecting the I-neg-space at all. (See axiom 4.) Similarly, the I-electrons (I-neg-mass) will fold and unfold with the I-neg-space, allowing the two streams to pass through each other with almost no interaction.

Theorem 23.

Individual I-electrons and individual I-positrons, in the type II unfolding mode, have nearly an identical reason for their interference patterns as that of I-light.

Proof:

> The interference pattern for an I-electron which exists in I-pos-space is associated with the folding and unfolding of I-neg-mass in I-neg-space. The I-positron exists in I-neg-space and its interference pattern is associated with folding and unfolding of I-

pos-mass in I-pos-space. There is interference when an individual electron is folding at the same place that a second individual electron is unfolding or an individual positron is forming. A similar statement holds for two individual positrons and an individual electron. I-light rays interference pattern occur under these same conditions.

Remark: Bombarding an I-atomic nucleus with an I-alpha particle (or any I-particle formed from I-swirls) as done by Rutherford, can interrupt the I-magnetism of the I-swirls in both the I-alpha particle and the I-atomic nucleus. This can create a short-lived I-north-pole to I-north-pole reaction or I-south-pole to I-south-pole reaction. In which case, an I-particle shoots out. The details are left open.

Theorem 24.

Let a body of I-mass have a velocity v where v is small compared to c. Let an I-light-ray be emitted either in the direction or the opposite direction of v from a source point on the moving body. Further, let this I-ray be reflected straight back toward the source point. When n cycles are used going out and n cycles are used coming back, the I-light-ray will, in both of the above cases, end up a little ahead of its starting point.

Proof:

The fact that in both cases slightly less dense pockets will be unfolded when going in the direction of v than when going in the opposite direction of v makes the two I-light-rays shift in the same direction. Recall, in the direction of motion the I-gravitational field is below the one over distance-squared type curve, but the density of pockets is increasing. In the opposite direction the I-gravitational field is above the one over distance squared type curve, but the density of the pockets is decreasing. Consequently, the two I-light-rays and the moving body will tend to all rendezvous after $2n$ cycles. Though in general all these changes are trivial compared to the speed of light, they do make it harder for Michelson and Morley's experiment to detect movement in I-space. (Moving gravitational fields account for most of the movement of the earth going around the sun when viewed from a housing Euclidean space. The speed through the ether is a very small percentage of the total moving speed. This, too, is hard on Michelson and Morley's experiment.)

Theorem 25.

I-atoms at a high velocity are more unstable than at a low velocity.

Proof:

As the velocity increases, by theorem 16 the I-gyros-1 open and use more and more I-space from outside the I-swirl, but use an invariant amount in the I-swirl for binding. With so many additional pockets being put through the flattened I-gyro-1s, pockets associated with the I-swirl become more and more disturbed and infiltrated. Thus, the I-atom becomes more unstable as it is controlled more and more by outside pockets rather than by those in the I-swirl.

Theorem 26.

I-electro-magnetic rays will lose energy when going outward in an I-gravitational field and will gain energy when going inward. The stronger the field, the greater is the gain or loss.

Proof:

For reason of simplicity, assume the I-electro-magnetic ray is going in the same or opposite direction as the density gradient in the I-gravitational field. When going outward, each folding phase will stop, according to axiom 5, part 8, when the density of the pockets at the folding chamber (the leading end) reach equilibrium with the surrounding pockets. But at that density, the pockets at the trailing end of the I-gauge-boson are not quite up to their surrounding pocket's densities. Thus, fewer pockets and, therefore, fewer nuggets, are folded than were unfolded. Consequently, energy or I-mass has been lost. Similarly, when the I-electro-magnetic ray is going in the same direction as the density gradient of the I-gravitational field, just the opposite is true. For this case, just as above, the folding phase will not stop until the density of the pockets at the folding chamber reaches that of the surrounding pockets, but now this means the pockets at the trailing end of the I-gauge-boson will exceed the density of the pockets surrounding them. Thus, in this case, additional I-mass or energy is produced.

Theorem 27.

A Black Hole in I-space will emit some radiation.

Proof:

By axiom 6, part 8 and part 10, an I-electro-magnetic wave unfolding perpendicular to the pocket's equal-density surfaces surrounding a Black Hole will continue to move outward, away from the Black Hole, until one of the two following things happen. (1) The I-electro-magnetic wave has sufficient energy to escape.

(2) The electro-magnetic wave does not have sufficient energy to escape. Then, by theorem 26, the I-electro-magnetic wave's energy will get so low that it goes below the critical value referred to in axiom 5, part 8. At this point, it fails to refold, and its stranded I-gauge-boson is I-heat. This I-heat or energy exists out some distance from the center of the Black Hole. A second I-electro-magnetic wave attaining this height can absorb this I-heat energy and go to a greater height. Repeating this process, I-heat is taken to higher and higher levels. Eventually, an I-electro-magnetic wave can pick up enough energy at enough different levels to completely escape the Black Hole. Note that the decreasing I-gravitational field at the higher levels means that a fixed amount of I-heat propels the I-electro-magnetic wave a greater distance as the height increases. This makes the total distance traveled have no upper bound. Consequently, given sufficient time, some I-electro-magnetic waves will escape. Also, the increasing I-heat at higher and higher levels means that entropy is increasing in the region of the Black Hole.

Note: Any forming I-photon that has equally dense pockets all around it is not bent back. Such an I-electro-magnetic wave occasionally escapes a Black Hole. However, near the center of a Black Hole, once an I-electro-magnetic wave starts crossing equal-density surfaces at any other angle than perpendicularly, it is bent back into the Black Hole (by theorem 8).

Theorem 28.

The outside surface of expanding I-space bends most radiation back into I-space.

Proof:

On the outside edge of I-space, there is a strong I-gravity field as described in axiom 3, part 6. Thus, by theorem 8, any I-electro-magnetic wave not crossing pocket's equal-density surfaces exactly perpendicularly will tend to be bent back into I-space. Note that this bent back radiation will make it appear that the outside edge of I-space is producing radiation. Also, any I-mass near the edge of I-space will tend to be pulled back into I-space by I-gravity. However, an I-electro-magnetic wave crossing the pocket's equal-density surfaces exactly perpendicularly will lose energy and will become I-heat, if it is not reflected back. If the I-heat forms at the very edge of I-space, it becomes part of empty Euclidean space. Therefore, at the edge of I-space, heat energy may cease to be energy, even though inside I-space heat is always potential energy.

Theorem 29.

It is possible to locate a line (really a circular parallelepiped) of uniform, dense I-mass buried in a uniform, less dense I-mass. (For example, it is possible to "witch" for a waterline buried in the ground.)

Proof:

In the following arguments, the small I-gravity of the involved rod is neglected.

Let a very long line of extra dense I-mass be completely removed from the presence of any other I-mass. Let a line, ℓ_1, be drawn perpendicular to the line of the dense I-mass at a point far from either of its ends. Mark a point A on the line ℓ_1 but away from the line of dense I-mass. Define the point A to be above the line of the dense I-mass. At the point A, construct a line ℓ_2 perpendicular to the plane of the above two lines. Let a rod of I-mass, with a very dense ball of I-mass fastened to its pointing end, be pointed toward the point A along the line ℓ_2. See figure 18.

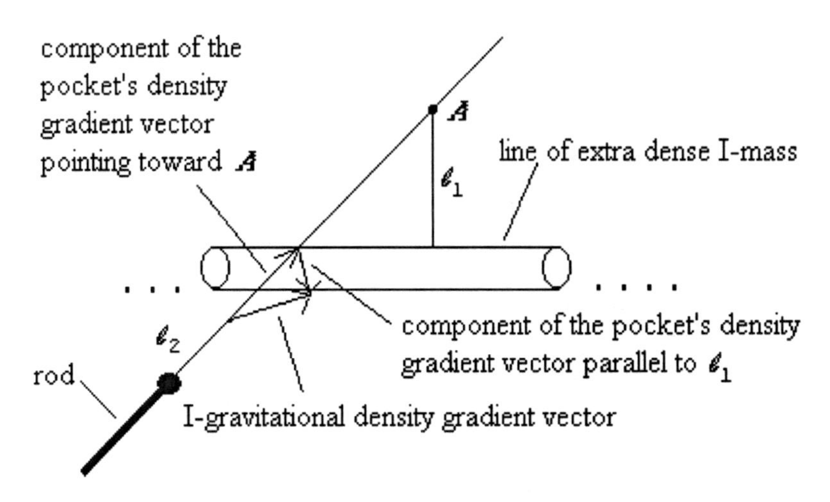

Figure 18. Witching rod setup

The pockets of I-space are compressed most at the surface of the line of dense I-mass, and there is perfect symmetry about each density gradient vector along the line of dense I-mass. Thus, the I-gravitational density gradient vectors, at any point outside the dense I-mass, are perpendicular to the line of dense I-mass. The components of the density gradient vectors, which are parallel to the line ℓ_2 will produce, an I-gravitational pull on the rod toward the point A. (Also, there are components of the density gradient vectors pulling the rod down. This pull can be overcome by forcing the rod to point toward A by using a swivel that leaves the rod free to rotate.) When the rod is carried past the point A, the I-gravitational pull will be reversed to again point the rod back toward the point A. This reversal occurs directly above the dense I-mass line.

Now take a large volume of less dense I-mass extending in all directions and leveled perpendicular to line ℓ_1 . Also, let the line ℓ_2 be above this leveled surface. When the line of dense I-mass is not there, the pocket's density gradient vectors associated with the less dense I-mass, by reason of symmetry, would all be perpendicular to both the leveled surface and the line ℓ_2. Thus, none of the I-gravitational vectors associated with the less dense I-mass will have a component pointing toward the point A. Consequently, these I-gravitational vectors will have no influence on the pointing of the rod toward the point A.

Next, remove a shaped hole of the less dense I-mass sufficient to allow the more dense line of I-mass to be buried. When the line of more-dense I-mass is put into this shaped hole, not only will it replace the less dense I-mass that was removed, but also will possess additional I-mass along the line. By axiom 3, part 7, at any point the gravitational density gradient vectors, from two sources, are additive. This additional mass will turn the rod as above. However, the turning forces will be reduced. The greater the difference in densities, the better the rod will work.

Remark: A similar proof would show that a line of less dense pockets of I-space could be found. The main difference in this case is that the rod will point away from the point A before and after passing the point. An example where a line of less dense pockets occurs is along a very intense beam of I-light. See theorem 13.

Theorem 30

Let L be a line in I-space along which the pocket's density gradient vectors are of the one over distance squared type. Also, let entropy be

increasing very slowly along this line. (In general, this would be a line starting at an interior point of I-space where the pocket's density has a maximal value, and then extending outward to the edge of I-space.) When relatively small bodies of I-mass are placed along this line, they all will move apart at an ever increasing rate, at least until one, or more, of the bodies reaches the point where the pocket's density has a maximal value. (Note that the bottom of a Black Hole is a point with such a maximal density. Also, note that to determine how the bodies are moving in a housing Euclidean space, one would have to take into account how fast the pockets along the line are expanding or contracting. If entropy is increasing very rapidly anywhere along the line, then in this region many pockets are unfolded and not replaced. This phenomenon will change the magnitude and direction of the involved pocket's density gradient vectors.)
Proof:

> The forces of I-gravity, between the relatively small bodies of I-mass along the line L, are neglected in the following argument. Along the line L, where the pocket's densities are of the one over the distance squared type, the magnitude of the pocket's density gradient vectors consistently decrease as the distance from the maximal point increases. This implies that each interior body of I-mass has an acceleration which is smaller than its neighbors that are closer to the point with a maximal pocket's density and has an acceleration which is greater than those that are farther away from this point. Thus, all of the bodies will be moving away from each other at an ever increasing rate, but all will be falling in toward the center of the universe.

Concluding Remark.

The author has tried to demonstrate that the structure of our universe is elementary, but by the same token it still possesses more basic particles and more basic independent properties than have been attributed to it over the centuries. The universe is understandable when the more complete structure of unitivity theory is used, but it appears that the limited structures that have been used over the centuries are too thin to carry a complete explanation of such things as light, gravity, magnetism, electricity, quarks, a unified field theory, etc.

The author encourages the reader to write and prove additional theorems that yield insight into various puzzling phenomena of the universe, to conduct experiments to check various implications of unitivity theory, and in the end to discover new truth that is implied by unitivity theory.

11 Dark-Energy and Unitivity

Let X represent the percent of the total energy of the universe that is composed of atoms, let Y represent the percent that is composed of cold-dark-matter, and let Z represent the percent that exists as dark-energy. Further assume that these percentages are given as integers with values between 0 and 100 with both end values included, and that X + Y + Z = 100. It is easy to show that there are 5,151 different ways X, Y, and Z can be chosen and still keep this equation true. The values for X, Y, and Z, have recently been measured very accurately. The values found are 4, 23, and 73 respectively.

Let us now investigate what unitivity theory has to say about these values. First the energy in atoms consists of nuggetrons; half of which are in the mass state and half of which are unfolded in the wave state. The force-carrying particles, called gluons, continually move nuggetrons from the mass state to the unfolded wave state and move unfolded nuggetrons back to their mass state with the aid of the folding-unfolding chambers. The rate and total time for the folding and the rate and total time for the unfolding of the gyro1s in any given quark are equal at all times. Thus, the total number of nuggetrons in the mass state is always invariant for any atom not moving in ether space. This coming and going of the gyro1s, which are completely out of sync, frees up some dark-energy pocketons for forming the ether space. Nuggetrons and pocketons come paired, and their energies cancel each other when they are folded or unfolded. Thus, the anti-energy of the associated, freed-up pocketons is equivalent to the total energy of the nuggetrons that are in the mass state and contained within the totality of atoms. This means an X/2 percentage of anti-energy is found in these freed-up and associated pocketons. This anti-energy is dark-energy. The X/2 nuggetrons in the wave phase contribute an amount of energy equivalent to that contributed by the nuggetrons in the mass state, and associated with these nuggetrons is a matched amount of dark anti-energy pocketons in their wave state. This gives the total percentage of energy that is associated with nuggetrons in atoms to be X. The total percentage of ant-energy in the pocketons which are associated with atoms is X also, and all anti-energy is dark energy.

The Y percentage of cold-dark-matter is again composed of nuggetrons because all matter is composed of nuggetrons. For example, the energy of a light ray can be in the state of the force-carrying particle, the photon, or it can be in the mass state as a gyro2, which in turn, is composed of nuggetrons and is always completely at rest in the ether space of the universe. Any ray spends an equal amount of time in each of these two states. This same type of event is true for all sub-atomic particles. They spend half of their time in a mass state and half of their time in a wave state as a force-carrying particle. The mass state of these sub-atomic particles makes up all of the cold-dark-matter. If there is a Y percentage of energy found in this cold-dark-matter, then there also is an equal Y percentage of dark-energy in the pocketons that are associated with this cold-dark-matter. Further, there is this same Y percentage of dark-energy in force-carrying particles quantized by first, the absence of nuggetrons and second, this same Y percentage by the absence of pocketons.

This now gives the total energy associated with the nuggetrons to be (X + 2Y) and the total energy associated with the pocketons to be (X + 2Y). Unitivity theory states that each of these must equal 50 percent. This gives the equation:

$$X + 2Y = 50$$

The total dark-energy percentage is found by adding (X + 2Y) from the pocketons plus Y from the absence of nuggetrons that are in a wave state and are associated with the cold-dark-matter. This gives the equation:

$$X + 3Y = Z$$

Again, the very accurately measured values are X=4, Y=23 and Z=73. Do these check out?

$$4 + 2\ (23) = 4 + 46 = 50$$

$$4 + 3\ (23) = 4 + 69 = 73$$

This is good news for both the people who made these measurements and for unitivity theory. There are only 26 groups of numbers that will satisfy all three of the above equations. Thus, that this is just a coincidence has the probability 26/5151 (or less). This value is approximately 0.005.

For discussion:

1. Why is the amount of energy in atoms such a small percentage?
2. What type of energy is heat? Hint: Heat is empty space.

12 Bouncing and Hit-Ball Mystery Solved

Introduction

Bouncing is a very interesting phenomenon. The complete story is a little lengthy, especially with the repetitions presented here to familiarize the reader with the details of bouncing. Thus, a condensed version is presented before the rest of the story in order to help the reader stay away from being unable to see the forest because of all the trees.

Where does one start to study bouncing? One thing that is very similar to bouncing is the change in directions of reflected light rays. Unitivity reveals that light in its mass form is a gyro2. A gyro2 is reflected or bounced back from a mirror in two different ways, and according to unitivity theory neither of these two ways affects the energy of the light ray. When a blue light ray is reflected, a cloned blue light ray bounces back. This property is true for light rays of any color. The fact that any light ray can be reflected in two different ways with one way losing a half wave length and the other way staying in phase, indicates that each light ray has two different ways of repeatedly being stored for a very brief instant of time. One way a light ray is stored is as a quantized, completely folded gyro2. The other way is as a quantized, shaped, empty hole which is formed in the ether space of the universe by the absence of a set of ether pocketons. This last form is the photon form and is obtained when the associated gyro2 is completely unfolded. In earlier chapters it has been shown that momentum is carried in gyro2s by the creative-destructive activity. Thus, when a light ray comes in contact with a mirror, the incoming momentum is imparted to either the mirror and/or the ether pocketons during the folding phase.

According to unitivity theory, in order to understand the bouncing of mass objects one needs to replace gyro2s with gyro1s, light waves with de Broglie waves, photons with gravitons or bosons, a mirror with a momentum field including its associated shockwave in the ether pocketons, and then continue to use the creative-destructive activity. This approach is described in the following paragraphs.

Newton's Laws

Momentum is carried in gyro1s by the creative-destructive activity just like it is carried in gyro2s. When any gyro1 is open and is in the folding phase, the creative-destructive activity pushes ahead on each folding nuggetron of the given gyro1 and, at the same time, pushes back precisely an equal amount on each folding pocketon. Newton's laws are always completely and precisely satisfied in unitivity, because the creative-destructive activity produces discrete movements that obey these laws exactly. For example, the force of gravity is produced and offset by the creative-destructive activity when, as stated above, it produces equal and opposite forces on nuggetrons and pocketons. Newton probably noticed that this type of phenomenon must be true because he consistently insisted that ether space must exist. The existence of gravity necessitates the existence of ether space in order to conserve momentum. A body falling in a gravitational field has an associated increase in the magnitude of its momentum, but without ether space, where is the offsetting momentum carried?

Magnitudes of Momentums in a Bounce

First, it will be established that it is theoretically possible to have an increase in the magnitude of the sum of the momentums of two colliding mass objects just after a bounce, when this magnitude is compared to the magnitude of the sum of their momentums just before the given bounce. This is basically like the increase in the magnitude of the difference between the momentums of two mass objects that are falling side-by-side in a given gravitational field. When the two, falling side-by-side objects have the same mass the magnitude of the difference is zero, but if one object has more mass than the other, then the magnitude of the difference in their momentums is an increasing function of time. It is this same type of an increase in the magnitude of the difference between two momentums that makes it possible to produce an increase in the magnitude of combined momentums during a bounce. Experimental results are included below to demonstrate that in some situations an increase in the magnitude of the sum of the momentums of two steel balls does take place when they bounce off one another. Given that one agrees with the results of this experiment, without an ether space where is the offsetting momentum for this increase in the magnitude of the combined momentums?

Unitivity theory implies that during the impact of one mass object hitting another mass object, the folding force associated with the creative-destructive activity forms a combined momentum field which when released yields an associated shockwave in the ether pocketons. The source point for a shockwave is the point or points where the two objects make contact. At this stage a huge number of things may happen depending on the speed, the materials, and the shapes of the incoming objects and/or the rebounding surface. In order to simplify the discussion presented here, only the bouncing of one steel ball off another larger steel ball will be discussed and checked by conducting experiments.

Bouncing Steel Balls

Unitivity theory reveals that the only way incoming momentum in one steel ball can be imparted to another steel ball is through a density gradient field that is created in the ether pocketons. This density gradient field is produced by the incoming steel ball's nuggetrons being pushed forward by the creative-destructive activity. These nuggetrons in turn push forward on adjacent local pocketons and/or adjacent local nuggetrons. If this pushing is of the "soft" pushing type, upon contact a combined density gradient field is formed in the ether pocketons that is referred to as a combined momentum field. This combined density gradient field causes any involved gyro1s to open and attempt to move in the direction of the local density gradient in the ether pocketons. In this situation after contact, the combined momentum field and eventually released shockwave may be too weak to produce very much bouncing back of the incoming steel balls. However, if the pushing is a "hard" impact type of pushing, then the combined momentum field is formed quicker and becomes stronger much faster. When such a momentum field is released, it takes the form of a shockwave in the adjusting ether pocketons. The combined momentum density gradient field and associated shockwave is dependent upon the magnitude of the incoming velocity and the amount of mass in each of the steel balls that hit one another. The velocity of each incoming steel ball can be quite small and yet the combined momentum field and associated shockwave may initiate a good bounce in which the gyro1s in one steel ball, or possibly in both steel balls, turn around and quickly accelerate away.

The Two Types of Bounces

In order for an incoming steel ball to bounce, the gyro1s in it must open in a reverse direction. Just as is the case for a gyro2, there are just two possible ways for a given gyro1 to turn around and go in an opposite direction. Each of these two ways is similar to one of the two ways that light is reflected, but each is slightly different, too.

In one type of bouncing, the gyro1s in an incoming ball must close completely as the ball comes to a complete halt, and then when the produced shockwave goes out past each of these gyro1s, they open in a reverse direction and move away from the point of contact. In this situation, each and every gyro1 in the incoming steel ball reverses directions in an at-rest, closed, mass state, exactly at a time when the gyro1 switches from folding in one direction to unfolding in a reverse direction. This will be referred to as a type-one bounce. The large number of times that the magnitude of the vector sum of momentums imparted to the two steel balls increased, or stayed about the same in the experimental runs conducted indicate that this is a very common way to produce a bounce. This type of bounce is studied in modern physics, and it has been likened to the loading of a spring and then releasing it. Unitivity theory reveals that the loaded spring is a momentum field and the released spring is the released shockwave which in turn is harnessed by the creative-destructive activity operating in the involved gyro1s. The change of directions for this bounce is just a little different than the corresponding change in directions for the reflection of light off the back side of a reflecting surface. The difference being that the incoming and outgoing speeds in a bounce need not be equal, but for this type of reflection of a light ray the two speeds are equal.

The second way a bounce may be obtained is for gyro1s in an incoming steel ball to remain open, and to change directions by folding the electron and positron in reverse positions at the opposite end of a formed graviton or boson. As will be proved a little later, a gyro1 reverses its direction of motion whenever its associated electron and positron are folded in switched positions with their spin orientations left unchanged. In this type of a bounce the gyro1s reverse direction similarly to the way gyro2s change directions when reflected from the front side of a reflecting surface. The incoming and outgoing speed of a ball making this type of a bounce can be nearly equal even before the added affect of the released shockwave further accelerates the outgoing ball. In this type of a bounce, gyro1s lose some incoming momentum that is switched into incoming momentum in the ether pocketons. In the reflection of light, gyro2s similarly lose incoming momentum that is turned into incoming momentum in the ether pocketons.

This bounce shall be referred to as a type-two bounce. In this bounce the gyro1s are forced to change directions by a strong combined momentum field and/or by its associated, strong, outgoing shockwave as the wave gets to them. This change in directions is in response to it being too difficult to fold gyro1s directly into very dense ether pocketons. Note that because some of the incoming gyro1s did not completely close, all remaining incoming momentum in such gyro1s has not been imparted to any thing, but on the first folding following the bounce this remaining incoming momentum is imparted to the local ether pocketons and consequently, it is mostly lost as far as mass objects are concerned. One must be sure to grasp the amazing fact that this remaining incoming momentum is converted into outgoing momentum in the involved ball, and it is free to mass objects in that all of this remaining momentum is transferred directly to ether pocketons.

Bouncing ball experiments were conducted to demonstrate that this type-two bounce truly does exist. A bounce in which the magnitude of the combined momentums has a "large" loss with no other explanation, demonstrates the existence of this second type of bouncing.

According to unitivity these are the only two ways that gyro1s can bounce. Thus, any bounce is explainable using some combination of these two ways of reversing the directions of gyro1s. Be sure to note, that in the bounce of any single steel ball, some of its involved gyro1s may use a type-one bounce, while other involved gyro1s may use a type-two bounce. Any object using two types of bounces possesses a great deal of strain, which is why some objects break in a bounce.

Combined Momentum Gained

In the universe, as given by unitivity, there is a source for gains in the magnitude of the combined momentums in mass objects that is acquired during a bounce. A large steel ball can harness a greater amount of momentum from a combined momentum field and associated shockwave than a small steel ball can harness in a similar situation. The reason being that each gyro1, at an equal distance from the source point of a given combined momentum field and its associated shockwave, harnesses approximately the same amount of momentum, and a large steel ball possesses many more gyro1s than a smaller steel ball. This completes the explanation of why in a bounce a larger steel ball can harness more momentum from the combined momentum field and its associated shockwave than is harnessed by a smaller steel ball.

It is true that when a large steel ball and a small steel ball are placed similarly near a combined momentum field and associated shockwave, the small steel ball's gyro1s "on-the-average" will harness a greater amount of momentum than the "on-the-average" amount of momentum harnessed by each gyro1 in the larger steel ball. This is due to the fact that a momentum field and associated shockwave weaken as the distance from the source point is increased. However, when everything is taken into consideration, the magnitude of the amount of momentum that is harnessed from a given momentum field and associated shockwave by a larger steel ball is greater than that harnessed similarly by a smaller steel ball.

As has been stated before, this same phenomenon is easily observed in a gravitational field where a larger steel ball with its greater amount of mass gains momentum (but not velocity) faster than a smaller steel ball gains momentum with its smaller amount of mass. Thus, in this situation, the difference of the magnitude of the momentums of these two steel balls with differing amounts of mass is an increasing function of time. The fact that this property is equally true in a combined momentum field and associated shockwave's density gradient field means that when only mass objects are considered, momentum may not be conserved in a bounce.

(Not that this is true only because one has left the ether pocketons out of the momentum calculations.) There will tend to be an increase in the magnitude of the combined momentums any time that both of the steel balls with their differing amount of mass are simultaneously harnessing some given density gradient field. The stronger the field and the longer the harnessing time, the greater is the increase in momentum. Given that a small steel ball produces all of a given shockwave, then a larger steel ball will generally harness the released shockwave more effectively than the small steel ball. In this type of a bounce at "lower" speeds there will tend to be a gain in the magnitude of the sum of the momentums of the two balls. At "higher" speeds the small steel ball can make a type-two bounce which then produces a loss in momentum that can overcome the gain.

Experimental Results

In each of four selected experimental bounces, a series of photos of two steel balls colliding were taken and the results are reported here to illustrate what happens during bounces. Each series of pictures makes it possible to approximate very accurately the momentum in each of the two steel balls just before and just after impact. For the first two described experiments, the small steel ball hits the large steel ball when it is basically at rest. In this situation and for the speeds photographed, the reflected shockwave produces a bounce-back momentum in the small steel ball which has a magnitude that is usually around 55% of the magnitude of its incoming momentum. The ratio of the mass of the large steel ball that was used to the mass of the small steel ball that was used is 3.63503822. The velocities are measured in pixels/sec and the unit of mass is taken to be the mass of the small steel ball.

Gains in the Magnitude of the Combined Momentums Illustrated

The combined incoming momentum for the run labeled *4coll*, is 11.8985 – 2281.045 = -2269.147 units (the large steel ball's momentum is given first), and the outgoing combined momentum is -3804.279 + 1318.189 = -2486.089 units. Thus, the shockwave moving through the two steel balls in this bounce produced a combined momentum having a magnitude approximately 216.943 units greater than the magnitude of the combined incoming momentums of the two steel balls. This means that there is about a 9.56% gain in the magnitude of the combined momentums of the two steel balls that is picked up during this bounce.

In the run labeled *9coll*, the combined incoming momentum for the two steel balls is 45.52651 - 843.1979 = -797.6717 units. The combined momentum after the bounce is -1408.624 + 463.1177 = -945.5063 units. Thus, in this particular bounce there is a gain in the magnitude of the combined momentum of approximately 147.8346 units, which is approximately an 18.5 % gain. Note, the large steel ball is basically at rest to start with in these two runs, but the small steel ball's incoming speed is much smaller in the second run. When this particular run, *9coll*, was repeated two more times with each of these runs having nearly an identical amount of input momentum, approximately the same percentage increase was obtained each time. The increase in the magnitude of the combined momentums in the two steel balls for these runs indicates that in these bounces, nearly all of the gyro1s in the small steel ball changed directions in a completely closed down, mass state. This is to say that most of the small steel ball's gyro1s had a type-one bounce. This has to be true in order to stay away from losing momentum. At higher amounts of input energy there was a smaller percentage gain, and sometimes a loss. This indicates that in a stronger momentum field, which is formed more quickly, some of the gyro1s in the incoming small steel ball use a type-two bounce. This in turn means some incoming momentum is lost to the mass objects because it has been converted into outgoing momentum.

Run 4coll

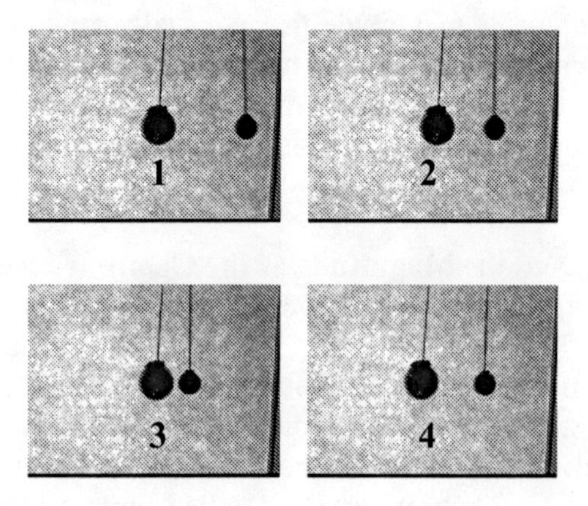

Run 9coll

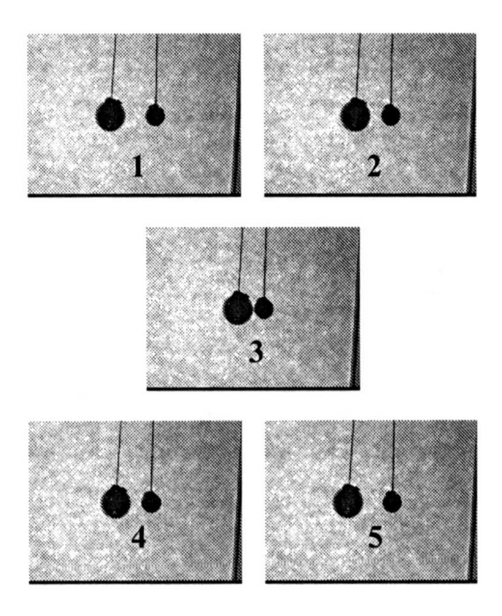

Losses in Combined Momentum Illustrated

There are situations where the magnitude of the combined momentums in the steel balls is decreased instead of increased during a bounce. One run of this type is the run labeled *coll.* In this run the large steel ball is coming in with 3116.7055 units of momentum and the small steel ball is coming in with -1035.971 units of momentum for a combined incoming momentum of 2080.7345 units. The combined outgoing momentum is1653.4028 units and is nearly all in the small steel ball because the large steel ball is nearly at rest with a momentum of -64.5612 units. The loss of approximately 427.3317 units in the combined momentum of the two steel balls means there is about a 20.54% loss in the magnitude of the combined momentum during this bounce.

Immediately, there are two questions that come to one's mind. How is it possible to have such a "large" loss in the magnitude of the combined momentums? How can a little steel ball carrying -1035.971 units of momentum stop a large steel ball carrying 3116.7055 units of momentum? It appears that the small steel ball must, somehow, get some help. Let us go step by step through this particular bounce to find the answers to these two questions.

Some stopping power of the small steel ball is lost unless its gyro1s keep going ahead until they are completely closed. During the collision of two steel balls, the combined momentum field grows and engulfs some of the involved gyro1s. When faced with a decrease in the density gradient field (but not the density field), gyro1s in the small steel ball begin to close down. If the momentum field becomes too intense, these gyro1s attempting to fold into the strong, combined momentum field are forced to turn around using a type-two bounce, and this event then releases the associated shockwave.

The small steel ball gets help stopping the large steel ball when the released shockwave reaches sets of out-of-sync gyro1s in the large steel ball which are switching from their wave-forming phase to their mass-forming phase and causes them to turn around in a type-two bounce. These turned-around sets of out-of-sync gyro1s in the large steel ball have their incoming momentum replaced with outgoing momentum. Thus, the gyro1s in these sets are now helping to stop the large steel ball by putting a backward force into it. On the other hand, the fast-moving shockwave cannot bring to a halt and turn around the sets of out-of-sync gyro1s that are switching from their mass-forming phase to their wave-forming phase. This means that the very fast-moving shockwave can cause about half of the involved, out-of-sync sets of gyro1s in the large steel ball's quarks to turn around, and must leave roughly the other half still coming in. Thus, the large steel ball ends up pushing nearly equally on ether pocketons in two, opposite directions, and consequently, brings itself to a nearly perfect halt. (This event allows one to view our ether space in action.)

On the other hand, the combined momentum field of the colliding steel balls remains in place long enough to cause the small steel ball's two out-of-sync sets in each of its quarks to either both have a type-one bounce, or both have a type-two bounce. When the small steel ball begins to move out, the combined momentum field begins to dissipate and move out as a shockwave in the ether pocketons. This shockwave passing through the small steel ball causes its gyro1s to open more which in turn makes the small steel ball accelerate. It then moves away at a high velocity.

It is very interesting the way that the universe can take slow-moving momentum in the large steel ball and convert it into fast-moving momentum in the small steel ball. To accomplish this feat in the small steel ball, the stationary combined momentum field converts the small ball's incoming momentum into outgoing momentum, and then the released associated shockwave speeds it up. In the large steel ball the shockwave turns around roughly half of the large steel ball's out-of-sync sets of gyro1s, and this in turn brings it nearly to a halt. This is exactly what the two balls need to do in order to mimic the conservation of momentum. To determine the exact momentums required to conserve both energy and momentum in this particular bounce, one can write down the two equations that must be satisfied, and then solve the quadratic equation that arises.

Now, the two stated questions can be answered. First, how is it possible to have a "large" loss in the magnitude of the combined momentums? In the run *coll*, the two steel balls colliding produce a strong combined momentum field which, when released, forms a fast-moving shockwave. This fast-moving shockwave passing through the large steel ball causes about half of the out-of-sync sets in the involved quarks to use a type-two bounce. This means that about half of the incoming momentum in the large steel ball is switched to outgoing momentum. This is where the loss in momentum comes from, and this is also the reason the little steel ball is capable of stopping the large steel ball even though the large steel ball may have much more incoming momentum. In short, at the quark level, the large steel ball uses the outgoing momentum of nearly half of the out-of-sync sets of gyro1s to cancel incoming momentum in the other nearly half of out-of-sync sets of gyro1s. Consequently, the large steel ball using ether space brings itself to a nearly perfect halt.

Note that there can be a gain, no change, or a loss in the magnitude of the combined momentums in the two steel balls during a bounce. The run *coll*, has a loss in the magnitude of the combined momentums, and the large steel ball ends up nearly at rest with -64.5612 units of momentum. In a similar run labeled 11b, the incoming velocity of the large steel ball is 1834.481 giving it a momentum of 6668.409 units, and the velocity and the momentum of the small steel ball are both equal to -2481.742 units. Thus, in the run 11b, the velocity for each of the steel balls is roughly 2.22 times those in run *coll*, which means there is roughly 4.93 times as much kinetic energy. In this run with the higher velocities the shockwave is strong enough to produce an outgoing momentum of 4428.233 units in the small steel ball, and the large steel ball is left nearly at a halt with only 17.91531 units of momentum. In this run there is a 6.2% increase in the magnitude of the combined momentums for the two balls. The 6.2% gain compared to the 20.54% loss in the magnitude of the combined momentums is quite large when one considers the fact that the two runs are basically identical except for the velocities involved. This shows how important kinetic energy is in a bounce. In this bounce with its strong shockwave, there is an indication that a large fraction of the gyro1s in the small steel ball, and about half of those in the large steel ball, had their incoming velocities converted over into outgoing velocities by using a type-two bounce. The type-two bounces in the small steel ball contribute to a gain, and the type-two bounces in the large steel ball contribute to a loss in the combined momentum after the bounce. Depending on which of these two is the larger, the total gain may be larger than the total loss or vice versa, and of course, the two could be equal.

Run Coll

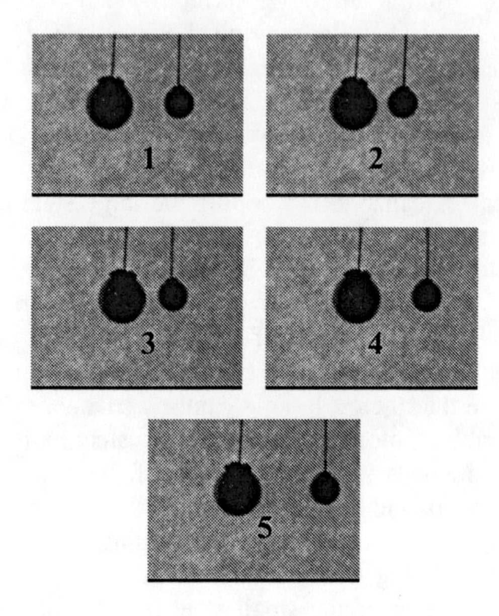

Note: 1035.971 units of momentum in the small steel ball halt 3116.672 units of momentum in the large steel ball. The "slow pitched" small ball departs with 1717.964 units of momentum which is an increase of 65.8% over its incoming units of momentum.

Run 11b

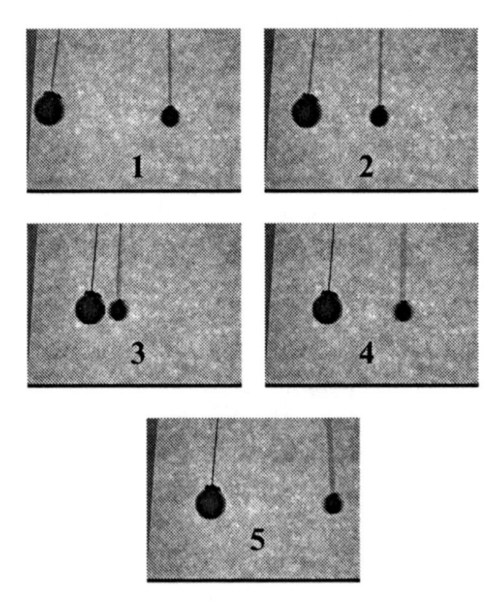

Note: 2481.742 units of momentum in the small steel ball halt 6668.409 units of momentum in the large steel ball. The "fast pitched" small ball departs with 4428.233 units of momentum which is an increase of 78.4% over its units of incoming momentum, and is 258% of the "slow pitched" ball's departing momentum, even though the large steel ball ends up nearly at rest in each of these two bounces. (Compare the two above pictures labeled 5, and also, see the Grand Finale Experiment at the end of this book.)

Brief Error Analysis

How reliable are these experimental results? The period between exposures is very consistent; down to less than a microsecond difference. Thus, this is not a significant source of error.

The use of a very short time interval for computing the velocities of the two steel balls just before impact and just after impact at a time when the two balls happen to be moving nearly on the level, means that these velocities increase only slightly during the interval right before the time of impact and decrease only slightly during the interval right after the time of impact. The way gravity affects these velocities is asymmetrical. Thus, the calculated velocities are nearly equally too small when they, respectively, are compared with the true incoming and the true outgoing velocities at the time of impact. When the magnitude of the combined momentums before impact is subtracted from the magnitude of the combined momentums after impact, these very small errors tend to cancel out making their contribution to the total error nearly zero.

When going at the same speed, the wind resistance on the large steel ball is slightly more than on the small steel ball. This tends to make it harder to obtain an increase in total momentum in the balls for a bounce in which the increase in total momentum is carried by the large steel ball.

The collision direction may not be exactly perpendicular to the camera's view angle, which would mean that the calculated velocities would be slightly lower than the true velocities. When this error exists it again makes it harder to obtain an increase in the magnitude of the combined momentums in the balls during such a bounce.

Inaccuracies in the location of each ball obtained using locating software produces errors in the momentum calculations. When these measurements are checked over and over using other means to obtain the location of the balls, the indications are that this error is less than one pixel per measurement, which translates to less than 1% error in the images obtained.

Momentum can be lost in rotational movement in the balls, and energy can be lost in the form of noise, heat, etc. These losses too, make it harder to obtain an increase in the magnitude of the combined momentums in the balls during a bounce.

Looking over this list of possible errors, and there really should not be any other major sources of error, it appears that the probability that errors alone can account for all the increases in the magnitude of the combined momentums of the two steel balls is essentially zero. Practically all the possible sources of error make it harder to obtain an increase in the magnitude of the combined momentums rather than easier. Thus, the gains in the magnitude of the combined momentums are difficult to question, and further, they are completely consistent with unitivity theory.

Going Back to Run *9coll*

Associated with the run *9coll* with its 18.5% gain and its – 843.1979 pixels/sec velocity are two other runs that were made which have about this same velocity on the incoming small steel balls and about this same percentage gain. A run with velocity -762.0388 pixels/sec had a 14.56% gain, and a run with velocity -866.8943 pixels/sec had a 16.47% gain. These gains are consistently quite large. This indicates that these gains are real.

In the cases where there are losses, the error analysis is not quite as supportive as in the case of gains. Having said this, the confirmed gains as given above, basically confirm the existence of offsetting ether pocketon's momentum, and strongly support the existence of combined momentum fields and their associated shockwaves in bouncing as is revealed by unitivity theory. In fact these gains support much of unitivity theory, and consequently support what unitivity has to say about losses, too. Some of the losses are really quite large, and they occur where unitivity theory indicates they should occur.

The Halting of the Large Steel Ball Reviewed

Besides percentage losses, there is another phenomenon that implies the existence of the type-two bounce. When the two experimental steel balls are dropped symmetrically from the same height so that their swings are affected equally by gravity, they will have extremely close to the same speed on impact. This means that the large steel ball has, in magnitude, 3.635 times as much incoming momentum as the small steel ball, and yet on collision at approximately the same speed, one always observes that the large steel ball comes to a nearly perfect halt. Even when one drops the two balls from slightly different heights, the large steel ball still tends to end up nearly halted. It appears that when a large steel ball's gyro1s are triggered into a bounce by any one of quite a range of momentums in the small steel ball, nearly half of its gyro1s perform a type-one bounce, and nearly half of its gyro1s perform a type-two bounce. When this is true, the momentums in the two sets of gyro1s use ether space to cancel one another, and the large ball brings itself to a halt.

Why is this most likely true? For one more time, according to unitivity, the story is the following. In each quark there are two identical sets of gyro1s which are completely out of sync. These out-of-sync sets can both have a type-one bounce, both have a type-two bounce, or the two out-of-sync sets can have opposite type bounces. When the two, out-of-sync sets have like bounces the quark bounces away. When the paired, out-of-sync sets in involved quarks have different type bounces, the involved quarks can end up nearly halted. For symmetrically dropped balls, because of a combined momentum field that is rather strong and that does not move out during the time it is being formed, nearly all the small steel ball's gyro1s in both out-of sync sets have time enough to bounce using the same type of bounce, and the small steel ball bounces away. While for most quarks in the large steel ball, the effect of the quickly formed combined momentum field does not reach out to them, nevertheless the associated, fast-moving shockwave that forms when this combined momentum field is released by the small steel ball moving out, does reach them. When the shockwave reaches a cluster of gyro1s, one of the out-of-sync sets in any involved quark bounces using a type-two bounce, and those in the other out-of sync set remain coming in and keep trying to make a type-one bounce. This means the momentums in the two, out-of-sync sets in the involved quark nullify each other, and this brings the involved quark to a halt. This last process, repeated many times over and going clear through the large steel ball, brings the tough, large steel ball nearly to a halt. (Note that a weak egg breaks in a bounce, but a tough steel ball does not break when stressed from within.) This is unitivity theory's explanation of the way the large steel ball comes to such a nearly perfect halt, and yet the small steel ball bounces away with a high velocity.

A large steel ball plows through a small steel ball that is at rest or has a very low incoming velocity, and in each of these situations it fails to halt. However, there are a host of combinations of velocities for the two steel balls that cause the large steel ball to nearly halt itself from within, while the small steel ball bounces away at a speed that is quite "fast."

Increased Forward Momentum Brought Back to a Nearly Perfect Halt

When a small steel ball imparts its forward momentum to a larger steel ball at rest, the large steel ball's forward momentum is greater than that carried by the small steel ball. For example in the run *9coll*, the incoming momentum is -843.1979 units in the small steel ball; the forward momentum imparted to the large steel ball is -1408.624 units, and the outward momentum on the small steel ball is 463.1177. When these balls are allowed to come back and make a second bounce, then in order for the large steel ball to return to its initial condition, (which needs to happen, if both momentum and energy are going to be precisely conserved), this large ball needs to have some sort of very accurate brake that is triggered by an incoming momentum of -463.1177 units in the small steel ball. Without such a brake in the large steel ball, the small steel ball has to muster up precisely 945.5063 units of outgoing momentum in order to bring the large steel ball back to a perfect halt. In magnitude this is more than the 843.1979 units of momentum that the small steel ball had when the process began. Yet, in spite of this apparent problem, after a second bounce the large steel ball is nearly back to its initial condition, and the small steel ball is only a little short of being back to its initial position. This is an indication that it is not the small steel ball that brings the large steel ball to a halt, but rather it is something going on in the large steel ball, itself, that brings it to a halt.

In the previously described run, *coll*, a momentum of 3116.705 units in the large steel ball is brought to a halt by the combined magnitudes of the incoming and outgoing momentums of 2753.935 units in the small steel ball. Without some sort of brake in the large steel ball, a halt under these conditions is not possible, yet a halt occurs. Thus, experimental results imply there has to be a braking mechanism in the large steel ball that brings momentum in the large steel ball to a nearly perfect halt when it is triggered by an incoming, small steel ball's momentum which, even along with its outgoing momentum, is too small to accomplish this feat. As stated before unitivity theory reveals that this necessary braking mechanism is due to type-two bounces canceling type-one bounces in approximately equal numbers within the large steel ball. Further, this braking confirms the existence of ether space.

Because of time considerations, only two balls were used in the experiments that were designed and run to establish the existence of the type-two bounce. It is highly probable that more favorable mass ratios can be found which would have even larger percentage losses and percentage gains. These experiments with their percentage losses and their percentage gains, along with the demonstrated, remarkable halting property, all agree with and support unitivity theory.

Ether Pocketon Momentum Reviewed

Unitivity theory shows that it is possible to have gains or losses in the magnitude of the combined momentums in the steel balls during a bounce, but when the ether pocketons are included then momentum is conserved. This is just the same as having an increase in the magnitude of the momentum of a mass object falling in a gravitational field, but when the ether pocketons are included in the calculations, the total momentum is conserved.

According to unitivity theory a bounce is produced by a fairly strong density gradient field in the dark energy pocketons that is in reality a combined momentum field, which when released becomes a shockwave. This means that for a perfect conservation of momentum in the balls themselves during a bounce, the combined momentum field and associated shockwave have to be formed and harnessed perfectly. This implies a higher accuracy than one would ordinarily expect in shockwaves. This is especially true when the two balls have different amounts of mass. This is where the ether pocketons become an absolute essential in order for our universe to maintain a precise conservation of momentum.

The Conservation of Momentum Reviewed

One last time let it be asked, what happens to the conservation of momentum in bouncing? Is it conserved or is it not conserved? Unitivity theory reveals that gravity employs a density gradient field to produce an observable difference in the magnitudes of the momentums of two, side-by-side falling mass objects which have different weights, and yet the conservation of momentum is not violated when the ether pocketons' momentums are included. Thus, when ether pocketons are included, it is true that the conservation of momentum holds in a bounce which also takes place in a density gradient field in the ether pocketons.

The creative-destructive activity pushes down on a falling mass object's nuggetrons and at the same time offsets these pushes with equal pushes up on ether pocketons. Thus, the conservation of momentum for a mass object falling in a gravitational field is attained by the creative-destructive activity employing ether pocketons. The true, precise conservation of momentum in bouncing is attained in exactly this same way. The only difference between the two cases is that in a fall the gyro1s open because of the local gravitational density gradient field; while in a bounce, the gyro1s open or reverse directions because of the local density gradient fields found in a combined momentum field and associated shockwave. Note that without the ether pocketons, the conservation of momentum is in trouble because when one considers just the mass objects alone in a fall, or in a bounce, momentum is not always conserved.

Using other words, gyro1s open in the direction of local density gradients in the ether pocketons. In the case of gravity the local density gradients are those of a gravitational field. For a bounce the local density gradients are produced by a combined momentum field and associated shockwave that are formed when one object hits another object. Individually and collectively, mass object's movements during falling and/or bouncing conserve momentum because the creative-destructive activity employs ether pocketons. This is a very precise conservation of momentum, and it must exist in order to keep the universe in balance. This means that the collective interactions between mass objects always conserve the true total momentum provided the ether pocketon's momentum is included. This is true in spite of the fact that the total combined momentum of observed mass objects themselves indicates that momentum is not always being conserved. A bounce or fall that creates mass momentum illustrates that there must be ether pocketons that are offsetting each and every momentum associated with moving mass objects. In reality, all momentums associated with a bounce or with a fall are conserved by the creative-destructive activity in exactly the same way that momentum is conserved in light.

Conservation of Energy

In addition to the conservation of momentum in a bounce there is the problem of the conservation of energy. In some of the bounces studied the sum of the kinetic energy in the two balls increases and in some bounces the kinetic energy decreases. In movement, according to unitivity theory, open gyro1s always convert kinetic energy in circuit pocketons to kinetic energy in the associated mass, and do this in an energy conserving way as is established in Chapter 10. Mass kinetic energy may increase in a bounce just like mass kinetic energy increases in a gravitational fall, but in each of these cases the total energy is always conserved when the circuit pocketons' kinetic energy is included along with the very important, necessary change in the unit of time. One can see the increase in the kinetic energy of mass that is produced by gravity or a bounce, but one needs the magic mirror of mathematics to see the offsetting decrease in the kinetic energy of the moving anti-mass in the two circuits of the associated quarks, and to see the required change in the unit of time.

The required change in time is the result of running the creative-destructive activity a little longer on each vibration in order to perform the additional work of removing pocketons from in front of the moving gyro1 and placing them behind the gyro1. The creative-destructive activity conserves both momentum and energy in the way it operates no matter how long it runs. But if one were to change the cross-sectional area of the two folding-unfolding chambers in a given gyro1 for the purpose of maintaining the same frequency of vibration and keeping time from slowing down whenever there is an increase in velocity, then one also would have increased the energy of the associated gyro1. This fact can be established by first changing a gyro1 into a gyro2, which from their construction, requires that the new gyro2 must be formed using the same electron and positron that made up the starting gyro1. Now increasing the cross-sectional area of the folding-unfolding chamber in the electron and in the positron of this newly formed gyro2 moves the energy of this gyro2 from the energy of say red light toward the energy of blue light. This means that any time the cross-sectional area of any given folding-unfolding chamber is changed there also is a change in mass energy. Consequently, the change in cross-sectional area in the situation as just described does not conserve energy. For more information on time see Chapter 10, theorem 16.

It is noteworthy that in the conservation of momentum, "ether" pocketons are required, and in the conservation of energy, "circuit" pocketons are required. This shows that in the conservation of momentum and in the conservation of energy, "ether" pocketons and "circuit" pocketons play different roles. They also play different roles when in the small, the "circuit" pocketons hold quarks and atoms together, and in the large, the "ether" pocketons hold stars and galaxies together. Note that the "circuit" and "ether" pocketons are identical and are just playing different roles.

Gyro1 in a type-two bounce

The type-two bounce will now be illustrated by a diagram. Consider the following projected gyro1 taken from a falling ball before bouncing:

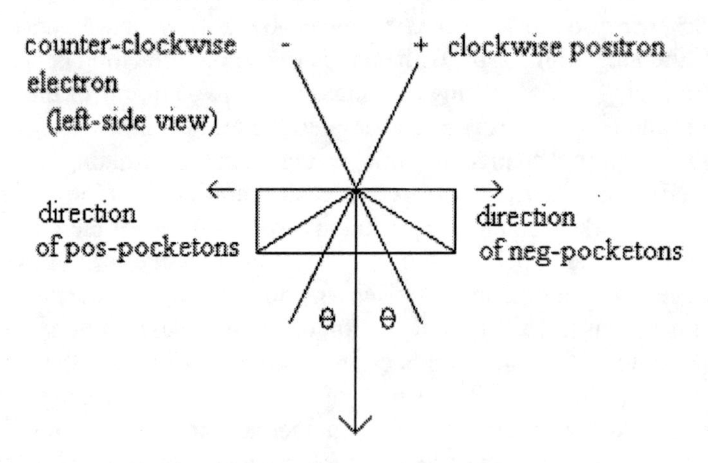

Projected gyro1 after bouncing:

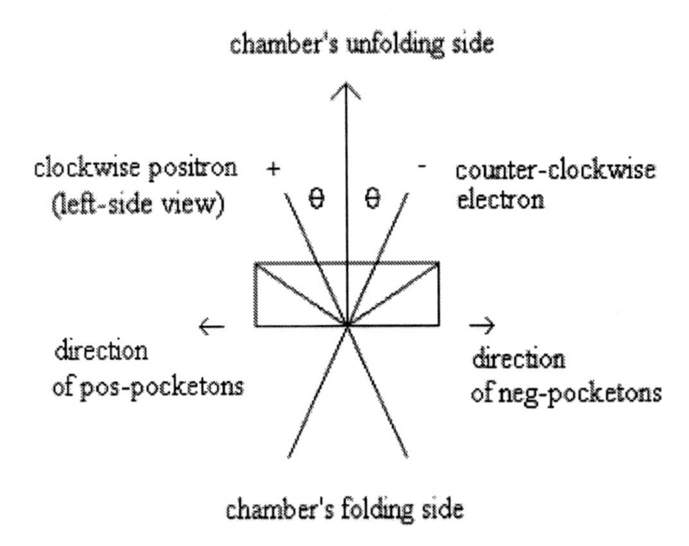

After a bounce the movement direction of the neg-pocketons and the pos-pocketons in their circuits must be the same as before the bounce. This is necessary in order to keep from destroying the involved quarks. This means that in a bounce either each involved gyro1 completely closes and then reopens going in a new direction, or this involved gyro1 remains open and its associated electron and positron are folded in exchanged positions in a single folding phase, but they both maintain their same rotation. This last situation is a type-two bounce and is illustrated in the above diagrams. After the bounce, the gyro1s open in a new direction and fold in a new direction as a result of the creative-destructive activity operating in the density gradient field that exists in the ether pocketons. For bouncing, the density gradient fields are a combined momentum field and an associated shockwave. Bouncing, just like gravity, requires ether space to carry the offsetting momentums for any and all of the mass momentums involved. The way this amazing feat is accomplished is reviewed below to make it absolutely sure that everyone gets it right. To those who got it the first time, this is boring, but it is better to be bored than to miss an important point.

Density of Pocketons and the Direction of Folding and Unfolding Reviewed

The change in the direction for unfolding or folding of a gyro1 is instigated by the increase in density of ether pocketons that occurs when the momentum field of one incoming steel ball collides with the momentum field of another incoming steel ball. On impact, the combined-momentum field forms a momentary, standing density gradient field that becomes a shockwave when this density gradient field is released. A gyro1 adjacent to an increase of density in the pocketons tends to unfold, and to open, in a direction toward the denser region, and to fold in a direction away from the denser region.

Shockwaves in Anti-Mass Reviewed

When pocketons are compressed at any point in space, these anti-mass particles produce an outgoing wave of rarefaction as the local pocketons pull adjacent pocketons in to support this compression. When the associated gyro1s in an impact region become closed or reverse directions in a type-two bounce, the force that created the compression is removed from the scene and the compression begins to dissipate. This produces an outgoing compression wave or shockwave in the sea of ether pocketons. When this shockwave passes by gyro1s, it causes gyro1s coming in toward the shockwave to change directions exactly as they switch from their wave-forming phase to their mass-forming phase, and it leaves coming in gyro1s switching from their mass-forming phase to their wave-forming phase modified, but still coming in. Gyro1s moving in the same direction as the shockwave momentarily open more after the shockwave passes as they attempt to follow it.

A compression wave in the ether pocketons can be observed as a shockwave producing movements of mass, but this wave is produced in, and is carried by, the expansions and contractions of the ether pocketons. This type of compression wave flips the gyro1s in a dribbled basketball making the ball bounce right back up to the dribbler's hand by using a combination of type-one and type-two bounces in the involved gyro1s.

These impact, dark-energy, density-rarefaction type waves that exit in the bouncing process are essentially gravitational waves. They are produced in the same way by the compressing of more nuggetrons and/or pocketons into a region, and they are carried in the same way by the expansions and contractions of ether pocketons. It is easy to observe dark-energy waves' effect on mass, but it is not easy to detect the dark-energy waves per se. Anyone seeking to detect gravitational waves would do well to study these identical, dark-energy shockwaves that are associated with bouncing. These impact, dark-energy waves or shockwaves are very local and very easily produced, and in addition, they are essentially gravitational waves.

Reversing the Direction of Momentum Reviewed

Normally, in a moving object, the folding and unfolding direction for the involved gyro1s is the same as the direction of motion. This is because of the way the creative-destructive activity moves mass objects through the ether pocketons, and also, at the same time produces effective forward momentum on the mass nuggetrons by imparting off-setting, backward momentum to the folding, anti-mass pocketons. In order for a ball to bounce and change the direction of its momentum, either its involved gyro1s must be forced to close and then unfold in a reverse direction, or its involved gyro1s must be in their wave form and then fold in a reverse direction. This means that in a bounce each involved gyro1 that reverses directions must have its unfolding direction opposite to its previous folding direction or its folding direction opposite to its previous unfolding direction for one vibration. A bounce is a reflection of momentum that takes place in gyro1s when they are completely in their mass state, or when they are completely in their wave state. This is basically the same phenomenon as the reflection of light which takes place in gyro2s that are completely in their mass state, or that are completely in their wave state. In short, it is reflected forward momentum that makes balls bounce.

Reflection of Light Reviewed

One way light can be reflected (perpendicularly) is to have the density of the pocketons around the leading end of the force-carrying particle, called the photon, sufficiently greater than the density of pocketons around the trailing end. A photon with its trailing end in air and its leading end in the glass of a mirror satisfies this condition. This situation causes the folding chambers in the associated gyro2 to flip their direction of folding and unfolding, and results in a returning ray that is half a wavelength out of phase when compared to the incoming ray. When reflection is completed the direction of movement for the light ray is reversed, and the direction for the backward push on forming ether pocketons is flipped, also. This means that the forward, effective momentum of the incoming ray is dissipated when the photon is formed and it is then replaced with reversed, outward, effective momentum in an identical, cloned, light ray. The offsetting momentum for the outgoing ray is in the ether pocketons, and some of this momentum may be transferred to mass objects because mass objects are encased in the ether pocketons.

Light can be reflected one other way by flipping the direction of the unfolding in gyro2s when the direction to the denser pocketons is opposite to the direction of motion. This is true because unfolding always tends to be in the direction of the density gradient in the ether pocketons. This type of reflection occurs on the backside of the glass in a mirror, and has all the same properties as the type of reflection that occurs on the front side of the mirror with the exception that the incoming and reversed, out-going rays remain in phase. After reflection, all the offsetting momentum of any light ray is as usual and is carried in ether space.

Type-two Bounces Convert Incoming Speed into Outgoing Speed in a Baseball

This discussion can all be summarized by saying, "In order to hit home runs, swing at fast pitched balls making sure the baseball is quickly and solidly exposed to a strong combined momentum field. A quickly-formed, strong, momentum field that is produced between the bat and the baseball forces the baseball to quickly reverse directions and to quickly accelerate outward. The fast incoming baseball helps one do this, because it contributes greatly to the combined momentum field, and this in turn forces the baseball to use a type-two bounce in which most of its incoming momentum is switched directly into outgoing momentum. The faster the baseball is coming in, the greater is the amount of this free, outgoing speed. To accomplish this one must be sure the bat does not bounce away from the ball, but rather be sure the baseball "clicks" away from the bat. The baseball, not the bat, must release the strong, combined momentum field. The bat must maintain good solid forward momentum and continue to have contact with the incoming ball for as long as possible in order to be sure that nearly all of the baseball's gyro1s are turned around in a type-two bounce before they can closed down. After this has been accomplished, the associated strong shockwave will supply additional outgoing acceleration to the hit baseball.

Or using other words, to hit home runs use a well designed bat that does not bounce away from a fast pitched baseball, but rather, quickly and very solidly exposes the baseball to a strong combined momentum field that makes the baseball "click" away from the bat. The mental goal of a good batter is that of quickly and solidly imparting a strong momentum field to the front of an incoming baseball. When this is accomplished, a type-two bounce takes place in the baseball wherein most of the baseballs gyro1s are forced to turn around in their wave form while they are still open. This means that the incoming velocity of the baseball is turned into outgoing velocity with only a relatively small percentage of its incoming momentum being imparted directly to the bat. When the quickly-formed, strong momentum field is released the associated, strong shockwave further accelerates the velocity of the outgoing baseball. The free, outgoing velocity contributed by the type-two bounce off a properly swung bat really helps the batter hit home runs.

The Introduced Mystery Is Solved

The mystery part of this book introduced in Chapter 1 is now solved. Gyro2s employ ether pocketon density to reflect the momentum of light from a properly prepared surface and do this in two different ways. Gyro1s employ ether pocketon density to reflect the momentum of bouncing balls, or of pitched balls, from a properly prepared surface and do this in two different ways. This means that a mirror and a baseball bat are in a sense alike. They both reflect mass.

A side remark is in order. If one could produce a compression wave or density gradient region in the ether pocketons that in turn could be used to control the direction of folding and unfolding of gyro1s by overriding the influence of a local gravitational field, then one could choose and maintain the direction of movement of these gyro1s. This would mean that one could fly at high speeds in the ether space of the universe; one could go wherever one wanted to go, and one possibly could accomplish this feat in a highly efficient manner.

Existence of an Ether Space Established

The fact that the magnitude of the momentum of a given mass object increases during a fall in a gravitational field, and/or during some bounces, implies that there has to be ether pocketons to carry all of the offsetting momentums; otherwise our universe would not always conserve momentum. Repeating for emphasis, an increase in the magnitude of the combined momentums of objects in a fall, or in a bounce, demands that there be ether space to carry any and all offsetting momentums.

Oh! To answer your important question, "Yes Virginia, there is a Santa Clause, and yes, there is ether space to carry his reindeer and his sleigh."

Henry Mattson, the author's father-in-law, repeatedly said, "Someday cars will travel without wheels, and the people that drive them will laugh at the people who thought they had to have wheels." The accomplishing of this momentous feat, obviously, is no small task, but unitivity theory does give a hint or two as to how this could possibly be done.

Gravity converts internal, potential energy within an object's quarks into external, visible kinetic energy and vice versa, and does this without using any energy in the process. Unitivity reveals the mechanism and principles that gravity uses to produce these perpetual accelerations. Now, one only needs to learn how to similarly produce, or at least "nearly" produce these same perpetual accelerations. Then cars can be made to travel without a need for wheels, and sleighs can be made to fly without a need for runners.

According to unitivity, bouncing uses the same mechanism and principles as gravity, but only for a brief period of time. Never-the-less, during this brief period of time, bounces can produce accelerations far greater than those produced locally by the earth's gravity. Thus, bouncing gives a hint as to what possibly can be done to produce unusually large accelerations. If a properly shaped and designed piston could be made to repeatedly strike ether pocketons solidly enough and quickly enough to cause the piston, itself, to make a type-two bounce off from the produced, density gradient field in the adjacent ether pocketons, the piston would bounce back out with more momentum than it hit with. This would be similar to the situation wherein a small steel ball flies out much faster than it comes in whenever it makes a type-two bounce away from the strong, combined momentum field that is produced when it collides with an incoming larger steel ball. The question is, "How can one replace or get around the contribution made by the incoming larger steel ball?"

Magnets move ether space, but in general they move positive pocketons and negative pocketons in opposite directions. If one could devise a scheme whereby the two types of pocketons could be moved essentially in the same direction, then one could move through ether space by removing the pocketons from in front of a given object and replacing them behind the given object. This would be like the way light moves through ether space. A light ray is moved through ether space by the creative-destructive activity destroying the pocketons immediately in front of the given ray when it is in its mass state and then reforming them behind the given ray. This means moving light is a perpetual motion machine.

Gyro1s are composed of electrons and positrons. If one could use an electric field somehow symmetrically or could use some other means to induce a given, combined set of gyro1s to simultaneously open in a predetermined direction then one could fly this combined set of gyro1s majestically through ether space.

The reader is challenged to think of other possibilities for making cars travel without wheels, and then proceed to check out the different possibilities in the laboratory.

Conclusion

Thank you for taking time to read about unitivity theory. Unitivity theory reveals a structure for our universe that indicates there are many fascinating unknowns just waiting to be discovered, and many strange mysteries just waiting to be solved. Good fortune to you as you attempt to unravel a few of the huge host of these unknowns and unsolved mysteries.

For discussion:

1. Discuss how a baseball bat and a mirror are related.
2. Discuss why bouncing implies the existence of ether.
3. Discuss the way a large steel ball determines its halting stance when it hits an incoming small steel ball.
4. Discuss the way refraction and the two reflections (bouncing) of light rays in rain drops produces two rainbows with one higher and backwards from the other. Hint: A reflection on the backside of a rain drop is refracted at least twice, and a reflection on the front side of the rain drop is refracted in an opposite manner, but only once.

Experiments for Checking Unitivity Theory

Rainbow and bouncing experiments will be discussed for those interested in doing work in the laboratory. There is no better way to checkout theory than to conduct experiments that are suggested by the theory. A couple example experiments of this type are given below.

The Two Rainbows Experiment

A question associated with bouncing is, "Why do light rays bouncing (or reflecting) off raindrops produce two rainbows?"

It has been proved in unitivity theory (and is well known in physics) that there are two ways for light to be reflected from any surface including raindrops. Light rays can be reflected from the front side, or the rays can enter a raindrop and then be reflected from the back side of the raindrop. Descartes uses the reflection from the back side of raindrops to show how the primary rainbow is formed by the sun's rays properly hitting the top portion of raindrops. It has been suggested that the way a secondary rainbow is produced is by "twice reflecting" in raindrops. Unitivity theory indicates that one should revisit and question this suggested way for creating a secondary rainbow.

According to unitivity, when a light ray enters a raindrop it does this by forming a photon which can be partly formed in air and partly formed in the water of the raindrop. It has been proved in unitivity that a photon being formed at an angle to a raindrop is always bent (curved) inward at the encountered point on the raindrop. The experiment introduced below shows that for a set of "close" together, nearly spherical raindrops exposed to a single source of light that hits their front side, there can develop one, and sometimes more than one, flare-up region or very bright region near the front surface of each involved raindrop. Smaller raindrops produce better flare-up regions than larger raindrops. A blue light ray's photon forming on the front of a raindrop is bent more than a red light ray's photon, and the amount of bending is a function of the angle between the entering light ray and the raindrop's curvature at the point of entry. The reflection or refraction of light off the front of the raindrop is produced by the folding chamber associated with the photon being forced back toward the air end of the photon rather than staying at the water end. This event reflects the light ray, and in addition produces refraction. When light rays have a common source, the reflected, less-bent, red light rays have a smaller angle of reflection than the angle of reflection for the more-bent blue light rays. This implies that when the spectrum is viewed from the same side that the light rays are coming from, the returning red rays appear closer to the incoming, source light rays than the returning blue rays. (It should be noted that this same refraction spectrum is produced when light rays are reflected back from the front side of snow crystals. The bright, sparkling spots that one sees in the snow contain the colors of the spectrum, and due to the refraction, they appear to be larger than they really are.)

The way this refraction is produced is as follows: The photon's folding chamber moving back to the surface of the raindrop, allows the reflected gyro2 to be formed in air rather than in water. The filling of a bent photon takes place very close to the surface of the given raindrop in the region where the photon is bent. This in turn produces two force vectors, one from the pushing of pocketons back into the portion of the photon that is surrounded by air, and the other from the pushing of pocketons back into the portion of the photon surrounded by water. These two force vectors determine the direction of motion for the reflected, emerging light ray. (Unitivity reveals that with this event there is a possibility for diffusion, a possibility for the production of heat, and associated with this heat a possibility for a change in the color of some rays.) But in general a less-bent, red light ray's photon is reflected back having a smaller angle of reflection than a more-bent, blue light ray's photon.

This event should be further researched, but to get started one can verify these statements to some extent by using what will be referred to as a T. J. Scmalkav machine. This machine is a clear bottle that is full of water with slightly distorted raindrops clinging to its outside at its spherical end which it should possess for doing this experiment. By rotating and moving this bottle while it is exposed to a single source of light (which for the eye's sake must not be too bright) it is possible to produce bright-light, flare-up points near the surface of some of the clinging raindrops. A small, local set of different raindrops hit similarly by incoming light rays will all develop small, flare-up regions in similar positions, and light rays having the same color and emerging from these associated, small, flare-up points will travel nearly parallel to one another.

In order to verify this statement, conduct the T. J. Scmalkav experiment as follows:

When a flare-up region is observed on the backside of the bottle, rotate it until the flare-up is elongated and the associated spectrum is displayed on the surface of the bottle along with the spectrum's mirror image. (This is accomplished by using only one eye, and to start with, looking nearly perpendicularly to the rays coming from the light source to the backside of the spherical end of the bottle. One may have to move one's head in order to follow a found, flare-up spot to a point where its spectrum is displayed.) The fire-bright spectrum when observed starts with red near the drop's bright spot and changes to blue as the distance from the drop's bright spot increases. The mirror image of this spectrum starts with blue and then goes to red as the distance from the focal point is increased. Also, "very-near-by" water bubbles are seen to be producing equivalent spectrums.

Now the way the secondary rainbow is formed becomes clear. The low, setting sun shining somewhat parallel to the earth's surface will shine on the lower half of some high raindrops. The rays on entering the lower half of one of these raindrops are bent upward. Any of these rays which happen to be reflected are sent back and downward forming spectrums. Note that reflected blue rays in this single refraction have a larger angle of reflection than red rays. And that the angle of reflection for each color is larger when reflected in a single refraction from the front of the raindrop than it is when the incoming rays hit the top side of a set of raindrops and are reflected in a double refraction from the back side of these raindrops. This is true because in double refractions the two different refractions tend to cancel each other. In fact, in double refractions blue rays end up having a smaller, final angle of reflection than red rays. Consequently, whenever reflected rays coming from raindrops are viewed by a single observer, reflected blue rays in the single-refraction event come from higher raindrops than reflected red rays come from, and the observed, single-refracted blue rays come from higher raindrops than the observed, double-refracted blue rays come from. According to unitivity theory, it is this single-reflection event taking place on the front side of the lower-half of raindrops that produces the secondary rainbow.

It is quite hard for the secondary rainbow to form and to make an appearance. Its associated "large" angles of reflection require very high raindrops, and the hitting of these raindrops on their lower half requires a sun very low in the sky. All of this is necessary, if the colored rays of the secondary rainbow are going to be seen by a close-in observer on the ground. Nevertheless, when all these conditions are met the secondary rainbow does appear with its color order backwards from the color order of the primary rainbow.

The secondary rainbow is usually not as bright as the primary rainbow. This is due to its being much higher and larger, and because, in general the percentage of the total number of incoming rays per square unit that are reflected from the front side of raindrops is smaller than the percentage that are reflected from the backside of raindrops. It should be noted that when the two rainbows exist, and the sun is setting, the primary rainbow tends to fade-away quicker than the secondary rainbow.

This completes an illustration of the way unitivity theory can assist a research scientist trying to understand a physical phenomenon of the universe. The brief discussion of the secondary rainbow presented here should really be studied further in the laboratory.

A Related T. J. Scmalkav Experiment

The T. J. Scmalkav experiment also, can be conducted on minute air bubbles clinging to the inside of the spherical end of the bottle in the same way it is conducted on the drops of water clinging to the outside. When a small air bubble on the backside of the bottle is exposed to light rays that hit the tangent plane to the spherical part of the bottle perpendicularly at the point where the air bubble is located, there are produced one or more flare-up points inside the air bubble. Such flare-up regions are formed from rays reflecting off the water surface of the involved air bubble, and this reflecting is similar to that of rays reflecting off water bubbles. However, now the curvature is concave instead of convex. This change in curvature changes the positioning of the flare-up regions, but the displayed, color spectrums in these two different cases are the same and go from a near-by red to a further-out blue. The existence of these fare-up regions with their spectrums verifies many properties of light as given by unitivity theory. This event, too, should be studied further in the laboratory in order to better understand the way refraction is related to reflection.

Please note that Unitivity theory, with its insistence that events and equations adhere completely to the structure wherein they are being studied, is a firm foundation for attacking unsolved mysteries of our universe.

The Grand Finale Experiment

College Physics by John A. Eldridge was the textbook used in the author's first course in physics at Montana State. Near the end of this book he relates, in so many words or less, that everything, including electrons, atoms, cars and people all move just as light moves; that is as waves, and only apparently as solid masses. Nothing moves or bounces exactly according to Newton's laws. Instead, everything moves according to a fundamental Wave Mechanics.

Unitivity theory is a fundamental Wave Mechanics. Unitivity is a theory that is halfway between relativity theory and quantum mechanics. In unitivity theory everything exists by alternating between a mass-forming phase, which is always at rest in ether space, and a wave-forming phase in which all movements through the ether space transpire. Further, the ether space of unitivity theory is dynamic and carries mass with it whenever and wherever it moves.

In unitivity, an at-rest quark in the sea of ether pocketons exerts pressure on its surrounding, adjacent pocketons producing a density-gradient, gravitational field. When this quark is also pushed into its adjacent ether space, an additional density gradient field develops in front of the quark, and the quark accelerates as its gyro1s open in the direction of the combined density gradient field. Having open gyro1s, the quark is capable of moving through ether space as a wave. When this pushing force is removed, all the involved gyro1s remain open, which means that during each of their mass-forming phases the gyro1s are pushed by the creative-destructive activity into the ether pocketons in front of them. This pushing repeatedly maintains the momentum density gradient field that was produced by the initial pushing. This momentum density gradient field is repeatedly maintained until it is acted upon by some outside force or is combined with some other field. Thus, a moving object that also happens to find itself in an overall gravitational field will be influenced by a single, combined density gradient field which is produced by the interaction of the object's momentum field and the adjacent overall gravitational field.

Now one might ask, "According to unitivity, how do quarks accelerate in a bounce?" When two steel balls collide, it is really their momentum fields that do the colliding because mass during its mass-forming phase is always at rest in ether space. In a collision, the two incoming momentum fields join together and produce a single, combined momentum field. This combined momentum field continues to get stronger until at least one of the incoming steel balls quits coming in. There are only two ways any ball can quit coming in. One way is for the ball's gyro1s to close down, and come to a halt. The other way is for the ball's gyro1s to turn around without completely closing down, and then start moving away. When one of the balls quits coming in, the momentum field is released and an associated shockwave in the ether pocketons begins to move out.

A bounce that occurs when the involved gyro1s close down and then open in an outgoing direction is a type-one bounce. A bounce that occurs when the involved gyro1s turn around, but never completely close down is a type-two bounce. Every bounce is either a type-one bounce, or a type-two bounce, or is some combination of these two types of bounces. All bounces are a function of the incoming momentum fields and the manner in which the resulting combined momentum field is released.

One can witness these two types of bounces employing the ether space of the universe by observing a large steel ball and a smaller steel ball bouncing off from one another. The following Scljamtkav machine helps one make these observations. (Actually, the accelerations of gravity display offsetting momentum in the ether space of the universe too, but in general it is not recognized.)

The Scljamtkav Machine

The directions for building a Scljamtkav machine are as follows: To make a rather small, carry-able machine, select a point on a circular wheel with a radius of about 3 inches (or if one prefers, of about 7.5 cm.) and trace out a curve by rolling the wheel along the top side of a straight line. The obtained curve is the well known cycloid which when inverted (turned upside down) forms the involute curve which has the tautochrone property. This means that in a gravitational field a pendulum swinging along this involute curve is isochronous. That is it takes the same amount of time to fall along the curve from any point on the curve to the curve's lowest point.

The pendulums to be used in this machine are two steel balls with different weights, each of which is suspended on a thread. The threads can be attached to the steel balls by using a welding-type, metal glue. The two steel balls referred to in the following experimental runs weigh 28.43 grams and 7.82 grams, respectively. The balls can be simultaneously released manually and observed physically. Or better yet, they can be simultaneously released electronically and observed by a fast-shudder camera.

To produce the evolute curve associated with the cycloid, take a fairly thick, rectangular piece of wood or some other similar material and cut out of its top a centered arc of the inverted cycloid. Take the top, inverted-cycloid portion and cut it into two parts by cutting perpendicularly to the curve at its thickest point. Use the right half for the left cheek, evolute curve, and the left half for the right cheek, evolute curve. Move these two cheeks without any rotation to the top of a second, larger rectangular piece of material and mount them there. The right and left cheeks are to be separated along the top line of the mounting board by a distance that is equal to the sum of the radii of the two steel balls that are used. Fasten a base under this second rectangular piece of material giving it the capability of standing alone. For the mounting and fastening, screws work nicely.

Take the first piece of wood and divide its bottom, inverted-cycloid portion into two parts by cutting perpendicularly to the lowest point on the curve. Mount these two parts with no rotation or switching of positions directly below their respective cheeks with the distance between these two parts being the same as the distance between the cheeks. Be sure to leave enough room below the cheeks so that the steel balls can swing freely. Using clamps suspend the each of the steel balls from the high point of its respective evolute curve. The thread length plus the radius of the hanging steel ball has to be equal to four times the radius of the circular wheel that was used to form the cycloid. In addition, these clamps can be used to fine tune the height of the two balls so that their centers are aligned.

For the releasing and the viewing of the bouncing steel balls, mark off along the mounted, bottom curve a set of swing heights in centimeters for the centers of each of the two steel balls. Use for the base or starting height the common, lowest height attained by the centers of the two, hanging steel balls. The heights can be obtained using either just a ruler, or better, using mathematics and the equation of the cycloid curve to obtain the desired x distance that is associated with each selected height value y. If a fast-shutter camera is used, all required distances and velocities can be obtained from photos. See the Scljamtkav machine pictured below.

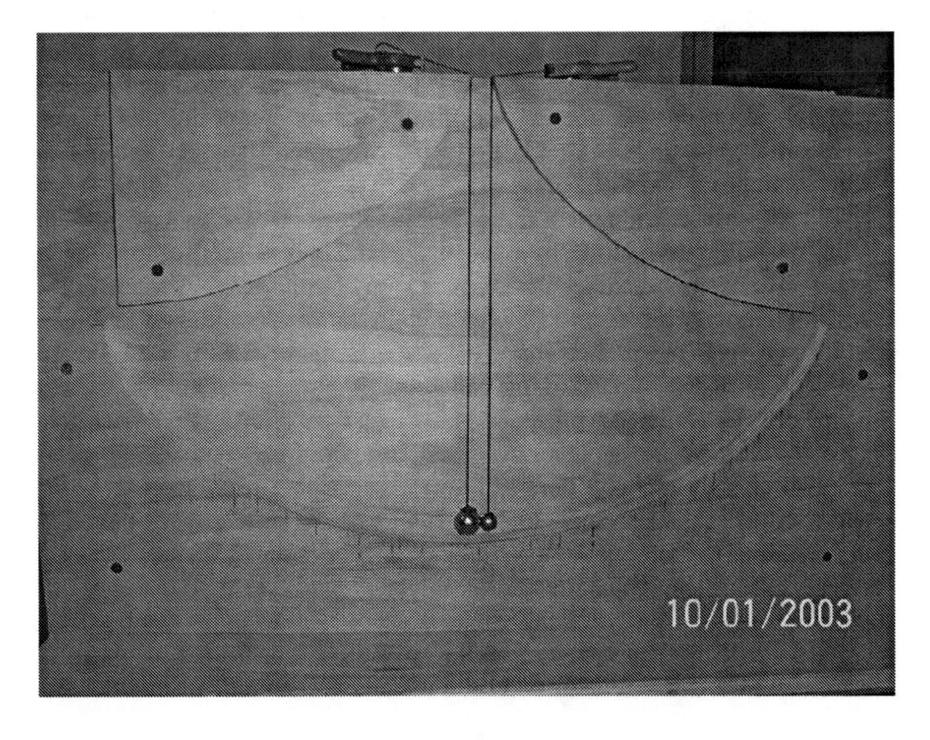

Knowing y, the perpendicular distance from the lowest point attained by a steel ball to the height at which it is released, one can compute the velocity v attained during its fall by using the formula v equals the square root of ($2 g y$). When the unit of length is centimeters and the unit of weight is grams, the gravitational constant g is approximately 980. The starting and ending height y for a steel ball can be determined by referring to the markings placed along the bottom curve. To determine the ending height, one can use the point where the ball just barely touches one's properly-placed finger. Given the weights of the two steel balls in grams, one can check the conservation of momentum in any given bounce of the two steel balls.

Experimental Runs

Run 1. Let the two balls just swing along together.

Notice that as the two balls swing back and forth they repeatedly start at rest and increase in velocity, then decrease in velocity and come back to rest. In order to conserve energy, physicists introduce potential energy. In order to conserve energy, the universe too, has to use potential energy which it converts into kinetic energy, and vice versa. According to unitivity, the creative-destructive activity is the mechanism that converts potential energy stored in the involved quark's circuit pocketons into the kinetic energy of the moving steel balls, and vice versa. (Recall that this stored potential energy is in the circuit pocketons and is equal to m times c-squared. The converting of this potential energy into kinetic energy requires a change in the unit of time that is given by the Lorentz transformations. All of this is derived in Chapter 10 where the opening and closing of gyro1s is discussed.)

In the case of momentum, the creative-destructive activity pushes and pulls on nuggetrons and simultaneously employs off-setting pushes and pulls on ether pocketons. This, according to unitivity, is the method by which the universe conserves momentum in the accelerations of gravity.

When watching the changing kinetic energy and the changing momentum of the two swinging balls, one is viewing the circuit and ether pocketons in action as they are being used by the creative-destructive activity to produce accelerations and decelerations that conserve both momentum and energy. One can also observe that friction in the strings and friction in the air cause the two steel balls to slow down. Given that the maximum velocity for the swinging balls is "small", it takes around 75 swings to lose about a centimeter of height.

Run 2. Simultaneously, drop the large steel ball from a height of two centimeters in the left wing and drop the small steel ball from a height of one centimeter in the right wing. Then repeat this event, but drop the small steel ball from a height of seven centimeters.

Notice that in both of these runs the two steel balls meet at the lowest point of the inverted cycloid, but in the second run the small steel ball bounces back and out much further than in the first run. The large steel ball possesses the same amount of incoming momentum for each of these hits. Consequently, these two different bounces illustrates that it really is much easier to hit a home run off a fast pitched ball than off a slow pitched ball.

In these bounces the small steel ball's incoming speed is changed to a greater outgoing speed. Unitivity points out that a large percentage of this outgoing speed is attained precisely at the time in the bounce when the colliding momentum fields for the two steel balls produce a combined momentum field that has become strong enough to force any and all gyro1s in the small steel ball that are still coming in to reverse directions when switching from their wave-forming phase to their mass-forming phase. The combined momentum field is not released until the small steel ball starts to move away. Thus, the combined momentum field exists long enough to force both sets of out-of-sync gyro1s within the incoming quarks of the small steel ball to reverse directions in an open mode. This event along with the boost from the released shockwave makes the little steel ball bounce away much faster than one would ordinarily expect.

Recall that a light ray is reflected from the front side of a reflecting surface when it turns around at the end of its wave-forming phase and exactly at the beginning of its mass-forming phase. A light ray is reflected from the backside of a reflecting surface when it turns around at the end of its mass-forming phase and exactly at the beginning of it wave-forming phase. In both of these reflections, the light ray is turned around with no change in color (i.e. no change in energy) and no change in speed (i.e. no change in the magnitude of the momentum).

The small steel ball's gyro1s can be turned around at the end of their wave-forming phase and exactly at the beginning of their mass-forming phase similarly to the way gyro2s in light are turned around when they are reflected from the front side of a reflecting surface. This type-two bounce leaves gyro1s in an open configuration, but their direction of motion has changed. During this type of bounce the small steel ball is turned around and starts leaving at nearly the speed at which it came into the bounce. In addition, it is free to harness the shockwave associated with the combined momentum field when it is released. In this type-two bounce, some incoming momentum is lost to the mass objects, because when gyro1s are turned around while open, their remaining incoming momentum is not transferred directly to any mass, but rather is converted over into outgoing momentum that has its off-setting momentum in the ether space of the universe.

In contrast, for a type-one bounce, gyro1s have to close and come to a halt in order to change directions at the end of one of their mass-forming phases and exactly at the beginning of the corresponding wave-forming phase. Thus, in a type-one bounce gyro1s are reflected back in a similar, but slightly different way than gyro2s are reflected back from the backside of a reflecting surface.

A very important observation associated with type-two bounces is that the acceleration obtained from the combined momentum field and associated, fast-moving-out shockwave is nearly instantaneous, and yet can be strong enough to greatly increase the height to which the small steel ball returns. Some of this return height is obtained by the open gyro1s being forced to change directions in the strong, combined momentum field, and the remaining portion is obtained by the gyro1s harnessing the outgoing shockwave. This means that the combined momentum field and associated shockwave along with the effect of a type-two bounce are much stronger than the earth's gravitational field. The obvious reason that this is true is that the earth's gravitational field requires quite a distance and quite a bit of time to bring to a halt the velocity that was imparted almost instantaneously in the bounce.

In "theoretical" bounces in which both momentum and energy are conserved for the above two runs, the small steel ball ends up at the heights, 7.8 centimeters, and 13.9 centimeters, respectively. For a T-ball situation where the small steel ball is similarly hit while it is at rest, "theoretically" it ends up at the height of 4.9 centimeters. This 4.9 cm. is only about 1/3 as high as the small steel ball goes when hit by the same swing in the case of the fast-pitched ball. The big question is, "How can the universe mimic these 'theoretical' bounces?" The answer is by using a combination of type-one and type-two bounces.

The remarkable Halting Property

The remarkable halting property of the large steel ball is observed in some bounces. The large steel ball is observed to halt, or at least very nearly halt, even though the magnitude of its incoming momentum is far greater than that of the incoming momentum in the small steel ball. In fact the large steel ball can be nearly halted even in a case where the combined magnitudes of incoming momentum and of outgoing momentum in the small steel ball is quite a bit less than the magnitude of the large steel ball's incoming momentum. This can be verified by using a camera version of the Scljamtkav machine. Also, it should be noted that impact, bouncing-away momentums of outgoing steel balls can be approximated by using the height to which each returns to compute the required velocity.

In various situations, why doesn't the large steel ball just keep going ahead, or just bounce away like the small steel ball bounces away instead of coming to a halt? Unitivity theory explains why the large ball tends to come to a nearly perfect halt under many quite different, initial conditions. Briefly the explanation is: Most of the two, associated out-of-sync sets of gyro1s that are found in each and every quark in the large steel ball end up going in opposite directions. Both of the out-of-sync sets of gyro1s for every quark close enough to the collision point have time to turn around, and produce a slightly distorted dented-in region, but the combined momentum field may not be strong enough to turn around quarks in the large steel ball that are too far away from the collision point. When the small steel ball begins to bounce away, the combined momentum field is released, and its associated shockwave starts to move out in all directions from the impact point. This fast-moving shockwave tends to force one of the out-of-sync sets in each of the out-lying quarks of the large steel ball to make a type-two bounce, but leaves the other out-of-sync set nearly alone. Just those out-of-sync sets that are ending their wave-forming phase and are starting their mass-forming phase get turned around with no loss in speed, while those in their mass-forming phase and starting to go into their wave-forming phase are modified, but are not even capable of being turned around instantaneously. This means the two, out-of-sync sets of gyro1s in each of the "too-far-away" quarks tend to end up going in opposite directions. The strong force circuit pocketons hold these quarks together, and the rugged, large steel ball does not break. Consequently, it is brought to a nearly perfect halt. (An egg would break.) When viewing the halting action in the large steel ball, one is viewing the ether space of the universe in action.

The universe does not produce a theoretically perfect bounce that conserves both energy and momentum. If the universe did produce such a perfect bounce, it would at the same time produce a perpetual motion machine. Nevertheless, the universe does mimic a perfect bounce, and at the same time does conserve both energy and momentum in these imperfect bounces. However, according to unitivity theory, the conservations take place at the quark level rather than at the steel ball level.

Run 3. Let the large steel ball be stationary at its lowest point on the involute curve. Next let the small steel ball fall along the involute curve from a height of one centimeter.

The theoretical, conservational, bounce-height value, for an at-rest, large steel ball of 28.43 grams that is hit by a given small steel ball of 7.82 grams dropped from a height of one centimeter, is approximately 0.19 centimeter. However, when this run is conducted over and over using the Scljamtkav machine, the large steel ball repeatedly ends up approximately at the height of 0.22 centimeters. (This observation can be obtained by using corresponding horizontal distances 3.4 cm and 3.65 cm respectively, and establishing that at 3.65 cm plus the radius of the large steel ball, the bounced, large steel ball just gets to and just touches one's finger properly placed at this point.) This means that the large steel ball gleans about 7.5 % more momentum from the small steel ball's momentum field and associated, moving-out shockwave than is required for the conservation of momentum. Again, the true conservation of momentum is in the quarks where mass interacts with ether space.

The small steel ball bounces back to about the height 0.3 cm. This is just slightly below the height of .323 which is the theoretical height required in order to conservation both energy and momentum.

At collision, the incoming momentum for the small steel ball moving to the left is 44.27 units, and the momentum for the stationary, large steel ball is zero. After the bounce, the large steel ball is moving to the left with 75.48 units of momentum, and the small steel ball is moving to the right with 24.25 units of momentum. Thus, after the bounce there is a combined momentum of 51.23 units moving to the left, and this means there is an increase in the momentum moving to the left of 6.96 units. This is a 15.7% increase, and is consistent with the high speed camera results previously obtained in similar situations like in the run *9coll.*

In the type-one bounce of this experiment, the small steel ball's gyro1s move ahead until they are forced to completely close down. At this point in time the momentum field is released, and the associated shockwave moves out through the two steel balls forcing gyro1s to open as it passes. The small steel ball's outgoing speed is much smaller than its incoming speed due to its incapacity to gain back everything it put into producing the combined momentum field. This is in contrast to a type-two bounce wherein the combined momentum field and associated shockwave both tend to be much stronger, and the two working together produce an outgoing speed in the small steel ball that usually is far greater than its incoming speed.

If one now lets the small steel ball bounce a second time, the second bounce is a type-two bounce and the two steel balls return nearly back to their initial conditions. But it is easy to see that the small ball does not get completely back. These observations indicate that some momentum has been lost to the ether in this sequence of two bounces. It is true that after the second bounce things are nearly back to their initial condition, but it would take two perfect bounces to bring each of the two steel balls back exactly to its initial condition. It is observed that the two bounces have taken more momentum out of the small steel ball than would have been removed by two, normal swings of the small steel ball.

When the two steel balls are allowed to repeatedly bounce off from one another, the small steel ball is observed to continually lose momentum. The large steel ball is observed to start with a very jerky, alternating type momentum wherein a larger in magnitude momentum is followed by a smaller in magnitude momentum and vice versa. Given a sufficient amount of time, the two balls converge to a smooth swinging momentum. In each type-one bounce that occurs, the large steel ball harnesses the shockwave associated with the combined momentum fields more effectively (This is true because it has more gyro1s.) than the small steel ball harnesses this same shockwave. Consequently, the large steel ball's momentum tends to increase from its initial, zero momentum, and the small steel balls momentum tends to decrease from its initial momentum wherein it possessed all of the momentum. This explains why the two balls having different masses end up swinging together, and they do not just come to some sort of a bouncing halt.

One should note that when the roles of the two steel balls are interchanged, the speed of the departing small steel ball greatly exceeds the speed of the incoming large steel ball. The large steel ball hitting the stationary, small steel ball causes the incoming momentum field to be focused and to become more intense in front of the small steel ball. This in turn causes the small steel ball's gyro1s to become more open than the gyro1s were in the incoming large steel ball. This means the small steel ball moves out at a velocity that is greater than the velocity of the incoming large steel ball.

Associated with this last observation, another interesting observation can be made by hanging a set of balls along a level, straight line such that each ball just touches its neighbor or neighbors and all of the ball's centers are at the same level. If the balls have increasing radii going from right to left, and if the small ball at the right end is allowed to hit squarely the next ball to its left in this lineup of balls, then a sequence of reactions takes place. These reactions produces a sequence of expanding momentum fields with each successive, larger ball harnessing its momentum field and associated shock wave better than does the adjoining, smaller ball that produced it. This means the left end ball picks up more momentum in this sequence situation than it picks up when it is hit directly by the small ball.

To verify this last statement, let the smallest ball at the right end directly impact the largest ball at the left end with no object between the two. Next, using the same velocity for the smallest ball let it hit a chosen sequence of say three inserted, increasing-in-size balls. When the two outcome momentums are compared, it is discovered that far less momentum is imparted to the largest ball in the case of a direct hit on the largest ball than in the case where the same sized hit goes through the sequence of touching balls which are increasing in size. The distance the largest ball goes in the sequence case is indeed greater than the distance it goes in the first case. This in turn, implies that the momentum obtained in the sequence case is greater than the momentum obtained in the direct-hit case. This is true in spite of the fact that the input momentum is the same for each case.

This phenomenon is interesting to watch, and the implications of the results are far reaching. This phenomenon definitely should be examined extensively in the laboratory.

The story of bouncing can be summed up by noting that the normal tendency is to think and explain bouncing in terms of mass and springs, but reality as revealed by experimental physics and unitivity demands that one think of bouncing in terms of ether fields and ether shock waves.

Printed in the United States
83697LV00004B/52-144/A

9 781425 992958